15

UNITED STATES MILITARY SADDLES
1812–1943

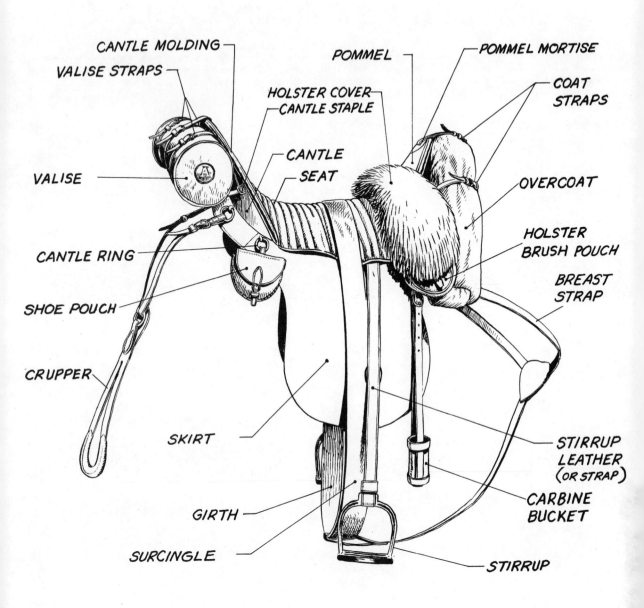

Nomenclature of the saddle. The saddle shown is the Model 1847 Grimsley and equipment.

UNITED STATES MILITARY SADDLES 1812–1943

by Randy Steffen

illustrations by Randy Steffen

UNIVERSITY OF OKLAHOMA PRESS
NORMAN

Washington, D.C.
12 April 1973

TO WHOM IT MAY CONCERN:

The Company of Military Historians, through its Review Board, takes pleasure in sponsoring *United States Military Saddles, 1812–1943*, by Randy Steffen, as an accurate and useful reference work in American military history.

George S. Pappas
President

Review Board
Harold L. Peterson
Carter Rila
Lee A. Wallace, Jr.
Brooke Nihart, *Editor in Chief*

By Randy Steffen
Horseman's Scrapbook, 3 vols. (Colorado Springs, 1959–65)
Horsemen Through Civilization, All (Colorado Springs, 1967)
United States Military Saddles (Norman, 1973)
The Horse Soldier, 1776–1943, 4 vols. (Norman, 1977–79)

Library of Congress Cataloging in Publication Data
Steffen, Randy, 1917–
  United States military saddles, 1812–1943.
  1. Saddlery—History. I. Title.
UC667.S76   357'.188   72–9268

Copyright © 1973 by the University of Oklahoma Press, Norman, Publishing Division of the University. Manufactured in the U.S.A. First edition, 1973; second printing, 1976; third printing, 1980.

PREFACE

From the time of Philip II of Macedon, who used cavalry troops in Europe three hundred years before Christ, until they became obsolete in World War II, horse soldiers were pre-eminent in warfare. Many nations have risen to greatness on the strength of their cavalry.

In any era three items have been indispensable to the cavalryman: his horse, his weapon, and his saddle. Without any of the three he was doomed to defeat or death. Millions of words have been written about the horses and weapons of conquest, but very little has appeared in print about the saddle—the equipment that delivered the horse soldier to the battlefield. This book represents a years-long effort to depict in words and illustrations the evolution of the saddle of the American cavalryman, who ranks among the greatest of the horse soldiers of history.

The saddles described and pictured in the following pages bore American cavalrymen in countries all over the world—in Europe, in the Orient, in Latin America—everywhere the United States displayed its military might. Horse soldiers crossed the Río Grande in the Mexican War, fought one another during the tragic, bloody years of the Civil War, charged up San Juan Hill in the Spanish-American War, defended American interests in China during the Boxer Rebellion, galloped across the hilly deserts of northern Mexico in pursuit of Pancho Villa, and charged across the muddy fields of France in World War I. They were an indispensable arm of American might until World War II, when the cavalry was converted to armor and the creak of saddle leather was replaced by the clank and roar of tanks and the odor of horse sweat was replaced by the stench of gasoline and oil.

Of all the achievements of the United States Cavalry, perhaps the most notable took place when its men, mounted upon sturdy horses, equipped with McClellan saddles and deadly carbines, surged across the western frontier and, in subjugating the Indians, assured the fulfillment of the American dream of manifest destiny—a country extending from coast to coast, a new power among nations.

History records comparatively little cavalry action in America's first wars, the Revolution and the War of 1812. Though the first official saddles were commissioned in 1792, mounted American troops played but a small role in those days. Mounted forces were then largely volunteers, civilians lacking government support or equipment. It is possible that military saddles were designed and contracted for at that time, but in more than ten years of intensive research I have been unable to discover any information about them or, indeed, concrete evidence of their existence.

After the War of 1812, the United States Congress became aware of the importance of mounted forces and began contracting for horse equipment. It is

PREFACE

those contracts, patents issued to servicemen and private saddlers, and other government documents on which this book is based. To the best of my knowledge every saddle issued as standard equipment by the United States Army is described and illustrated in the pages that follow. The saddles, uniforms, and other equipment were drawn from actual specimens in my own and other private and public collections or from descriptions in official records in the National Archives and War Department records. The sources of my research are indicated in the text or in the legends accompanying the illustrations.

The evolution of the American military saddle is a story of continuing efforts to improve upon existing equipment, often hampered by a parsimonious Congress or peacetime indifference. Over the years many innovations were devised to cope with the horses the soldiers rode, the terrain they had to cross, and the weather they had to endure. Many famous military men took part in the effort to produce the "ideal" saddle, among them George B. McClellan, William Tecumseh Sherman, George Armstrong Custer, and John ("Blackjack") Pershing. These efforts were surrounded by rumor and gossip, much of which has persisted to this day. One example is the curious claim of General McClellan that his famous and often-revamped saddle was modeled after saddles he studied in Europe. Most military historians accept without question this myth, which is cleared up in Chapter 5.

Despite the failures and disappointing designs, the American horse soldier was probably the best-equipped cavalryman in history. Latter-day saddles, modified McClellan and Phillips patterns, are still in use, notably in North America and in Australia, a fact which testifies not only to their durability but also to their continuing usefulness in an increasingly mechanized world.

This book is in part the result of a lifelong interest in horses and horse gear. Those interests were reinforced by a year at Stanton Preparatory Academy and four years at the United States Naval Academy. For nearly thirty years there have been few days when my activities did not center around horses and horse gear of one kind or another.

In the 1940's I began writing and illustrating articles on various aspects of horsemanship. I realized how important it was to study old-time horse equipment, for most of my articles dealt with the horses, stockmen, soldiers, and Indians of the nineteenth century—those colorful actors in what was to me the most absorbing period of our history.

Membership in the Company of Military Historians sparked a deepening interest in early American military history. Upon learning that little material was available on early-day horse equipment, I vowed that one day I would write a book on military saddles.

Having a hard-headed attitude about authenticity in illustration details, I began gathering specimens of the equipment called for in military illustrations. Over the years these "props" grew into a large collection requiring a sizable studio. In 1963 a fire broke out in the studio, destroying all of the collection. The treasured military artifacts that I had jammed into that two-story log building were destroyed— hundreds of guns, saddles, bits, spurs, bridles, uniforms, equipment, and accouterments, as well as Indian weapons and clothing, not to mention a large

PREFACE

reference library and thousands of photographs and notes. It was a devastating blow, but, thanks to many good friends and much good fortune, I have managed to build another collection, which has guided me in my efforts to complete my research.

I am indebted to many persons for their help in seeing this book to publication. I gratefully acknowledge the assistance and encouragement of the personnel of many government agencies throughout the United States. To list all of them would require many pages. I am especially grateful to Harold A. Geer, of Ravenna, Ohio, for unselfishly sharing his collection and knowledge at a time when I was hampered by the loss of my collection and notes and for his constant encouragement in later years. To many others I owe a debt of gratitude I can never repay. Finally, I express my appreciation to my wife, Dorothy, without whose patience and constant help this book would never have been completed.

Some of the material in the book was previously published in a series of articles in the *Western Horseman*, which graciously granted me permission to adapt it for this book.

It is my earnest hope that the information in the pages that follow will be a genuine contribution to the historians and artists of America's horse soldiers.

Walking S Ranch
Dublin, Texas

Randy Steffen

CONTENTS

		Page
	Preface	v
1	The Walker Contract Saddle and Other Dragoon Saddles, 1812–14	3
2	Dragoon Saddles, 1833–41	17
3	The Ringgold and Grimsley Saddles, 1844–47	33
4	The Campbell, Hope (Texas), and Jones Saddles, 1855–58	49
5	The McClellan Saddle, 1859	63
6	The Jenifer Saddle, 1860	75
7	Modifications of the McClellan Saddle and the Whitman Saddle, 1868–1904	79
8	Cavalry Board Saddles, 1912–16	107
9	Officers' Saddles, 1917–26	121
10	The Last McClellan and the Phillips Saddle, 1928–43	131
11	Miscellaneous Saddles	141
	Index	156

ILLUSTRATIONS

		Nomenclature of the saddle	Frontispiece
Fig.	1	Walker saddle, 1812	Page 5
Fig.	2	Sergeant, First Regiment, Light Dragoons, with Walker saddle, ca. 1812	7
Fig.	3	Private, Second Regiment, Light Dragoons, with Walker saddle, ca. 1812	8
Fig.	4	Hussar-style saddle used by some Light Dragoons, ca. 1812	11
Fig.	5	Lieutenant colonel, First Regiment, Light Dragoons, with hussar-style saddle, ca. 1812	12
Fig.	6	Issue dragoon saddle and pad, ca. 1814	13
Fig.	7	Captain, Second Regiment, Light Dragoons, with issue dragoon saddle, ca. 1812	14
Fig.	8	Model 1833 dragoon saddle	18
Fig.	9	Sergeant, Regiment of Dragoons, with Model 1833 equipment, ca. 1837	20
Fig.	10	Private, Regiment of Dragoons, with 1833 equipments, ca. 1837	21

Fig. 11	Corporal, Regiment of Dragoons, with 1833 equipments, ca. 1834	23
Fig. 12	Lieutenant, regimental staff, Regiment of Dragoons, with 1833 dragoon saddle, ca. 1834	25
Fig. 13	Model 1841 dragoon saddle	27
Fig. 14	Private, Second Regiment of Dragoons, with Model 1841 equipments, ca. 1842	29
Fig. 15	Sergeant, Second Regiment of Dragoons, with 1841 equipments, ca. 1842	30
Fig. 16	Ringgold dragoon saddle, 1844	35
Fig. 17	Private, Second Dragoons, with Ringgold equipments, ca. 1846	39
Fig. 18	Private, Second Dragoons, with Ringgold equipments, ca. 1846	40
Fig. 19	Grimsley dragoon saddle, 1847	42
Fig. 20	Corporal, Regiment of Mounted Rifles, with Grimsley dragoon saddle, ca. 1846	45
Fig. 21	1847 Grimsley dragoon officer horse equipments	47
Fig. 22	Campbell saddle, 1855	52
Fig. 23	Private, First Cavalry, with Campbell equipments, ca. 1855	55
Fig. 24	First sergeant, Second Cavalry, with Campbell equipments, ca. 1855	56
Fig. 25	Hope saddle owned by General Joseph E. Johnston	58
Fig. 26	Hope saddle owned by General Hugh Judson Kilpatrick	59
Fig. 27	Hope saddle, ca. 1856	60
Fig. 28	Jones steel-frame adjustable saddletree, 1854	61
Fig. 29	Model 1859 saddle (first McClellan)	67

ILLUSTRATIONS

Fig. 30	Sergeant, First Cavalry, with Model 1859 equipments, ca. 1864	68
Fig. 31	Corporal, First Cavalry, with Model 1859 equipments, ca. 1864	71
Fig. 32	Modified McClellan officers' saddles, 1860–75	72
Fig. 33	Jenifer cavalry saddle, 1860	77
Fig. 34	Modifications of the McClellan saddle	83
Fig. 35	Captain, Third Cavalry, ca. 1875	84
Fig. 36	Private, Third Cavalry, ca. 1875	86
Fig. 37	Sergeant major, Third Cavalry, ca. 1875	87
Fig. 38	Model 1874 McClellan saddle	88
Fig. 39	Captain, Third Cavalry, with Model 1874 equipments	90
Fig. 40	First sergeant, Third Cavalry, with Model 1874 equipments, ca. 1878	91
Fig. 41	Corporal, Third Cavalry, with Model 1874 equipments	92
Fig. 42	Whitman cavalry saddle, 1879	95
Fig. 43	Whitman saddle without horn, 1879	96
Fig. 44	Private, Cavalry, with Whitman equipments, ca. 1881	98
Fig. 45	Model 1885 McClellan saddle	100
Fig. 46	Sergeant major, Sixth Cavalry, ca. 1885	101
Fig. 47	Private, Sixth Cavalry, ca. 1885	102
Fig. 48	Model 1904 McClellan saddle	103
Fig. 49	Private and sergeant of cavalry with Model 1904 McClellan saddles, ca. 1912	109
Fig. 50	Model 1912 cavalry saddle	111
Fig. 51	First sergeant and private with Model 1912 equipments, ca. 1912	113

Fig. 52	Captain, Fifth Cavalry, with Model 1912 equipments, ca. 1914	114
Fig. 53	Captain, Fifth Cavalry, with Model 1912 equipments for garrison duty, ca. 1914	116
Fig. 54	Model 1916 cavalry saddle	119
Fig. 55	Model 1917 officer's field saddle (Ordnance Model 1916)	122
Fig. 56	Captain, Second Cavalry, with Model 1917 officer's field saddle, ca. 1917	123
Fig. 57	Company officer, Cavalry, with Model 1917 officer's field saddle, ca. 1931	124
Fig. 58	Model 1916 officer's training saddle	125
Fig. 59	Corporal, Second Cavalry, with Model 1904 saddle, ca. 1917	126
Fig. 60	First sergeant, Third Cavalry, with Model 1904 saddle, ca. 1917	127
Fig. 61	Model 1926 training saddle, French Saumur type	128
Fig. 62	Model 1928 McClellan saddle	132
Fig. 63	Private, Seventh Cavalry, with Model 1928 McClellan saddle, ca. 1931	133
Fig. 64	Private, Seventh Cavalry, with Model 1928 McClellan saddle, ca. 1931	134
Fig. 65	Private, Fourteenth Cavalry, with Model 1928 McClellan saddle, ca. 1941	136
Fig. 66	Model 1936 Phillips officer's cross-country saddle	139
Fig. 67	Model 1832 artillery driver's saddle, and Model 1832 artillery valise saddle and valise	143
Fig. 68	Corporal of Light Artillery, with 1832 driver's saddle	144

ILLUSTRATIONS

Fig. 69	Model 1859 Grimsley artillery driver's saddle and artillery valise saddle	147
Fig. 70	Driver's saddle for horse and mule jerk-line teams, used from 1959 through the 1880's	149
Fig. 71	Wheel team of a six-mule jerk-line team hitched to an army supply wagon, used from the 1850's through the turn of the century	151
Fig. 72	Model 1913 mule riding saddle	152
Fig. 73	Full-rigged packer's riding saddle, ca. 1917	153
Fig. 74	Skeleton-rigged packer's riding saddle, ca. 1917	154

UNITED STATES MILITARY SADDLES
1812–1943

1 THE WALKER CONTRACT SADDLE AND OTHER DRAGOON SADDLES, 1812-14

The first saddles manufactured for and issued to United States mounted troops may have been those made for the one company of dragoons authorized by Congress in 1792, more than ten years after Cornwallis surrendered in 1781. But no records of standard horse equipments for that handful of cavalry seem to exist. There is little doubt that the two companies of dragoons, to which the mounted arm was increased in 1796, and the full regiment authorized two years later were issued standard horse gear. But in more than ten years of intensive research I could find no records indicating that such equipment was contracted for.

The first reference to saddles and bridles for dragoons that appears in the records of the National Archives, Washington, D.C., is the contract issued to James Walker, a Philadelphia saddler, by Tench

Coxe, purveyor of public supplies. This contract, dated May 9, 1812, was for

> 200 sets of horse equipments, not inferior to the patterns, and fit for effective military service, as follows, viz.: One strong horseman's or trooper's saddle, lined with stout, undyed, twilled cotton serge, stuffed with hair, with a brass head and cantle, improved stirrup irons, two girths (worsted webbing therefor to be furnished by the United States); One leathern breast plate, iron rings and staples to saddles; One valise pad with straps to secure it to saddle; One valise of Ravens [?] Duck painted with fixed leathern buckle straps, loop straps, and straps to secure it on the pad; One trooper's leathern halter; One double bridle bit and bridoon; One pair of holsters; One pair of holster covers, covers of bear skin; One leathern circingle of bridle leather.

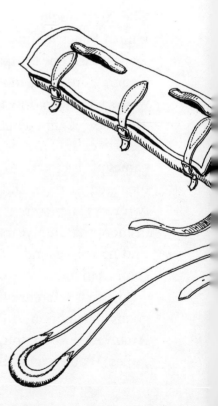

THE WALKER AND OTHER EARLY SADDLES

Figure 1 shows this saddle in detail, with the quilted seat and valise pad detached so that the details of the hammock-type seat and the attachment straps of the valise pad can be seen.

The use of a crupper, shown attached in Figure 1, was left to the judgment of the troop commander when saddles with double girths were provided. A crupper is shown attached here to illustrate how it was threaded through the loop on top of the valise pad. Military men of that era considered it important

Fig. 1. The saddle designed by James Walker, a Philadelphia saddler, in 1812.

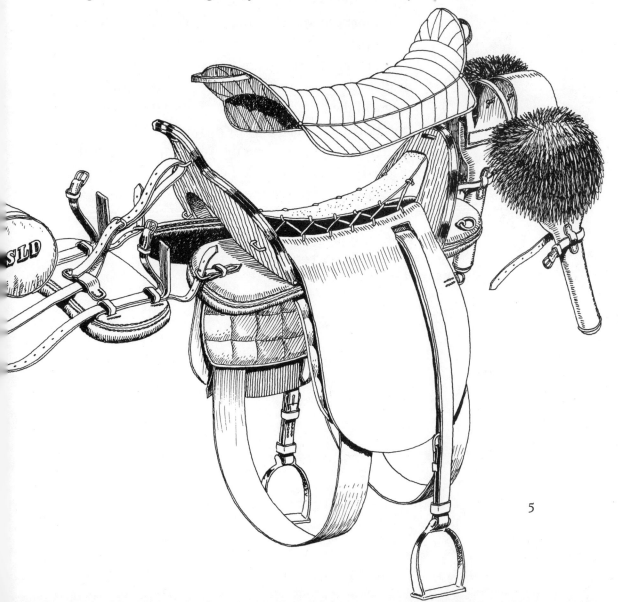

that no part of the valise come into contact with the horse's back—thus the use of a pad, or postilion, until the equipment change in 1841. The theory was that the valise, securely fastened to the pad, which rode directly on the horse's back behind the rear of the saddle bars, did not move on the back to cause friction sores. The pad, though fastened to the saddle with its straps, was completely independent of the saddle and its flexing on the horse's back. Later the valise and other cantle packs were fastened directly to the cantle high enough to clear the back, so that the pad was no longer needed.

As one can see in Figure 1, this saddle is a simple three-piece one, the pommel and cantle being mortised or otherwise securely fastened to the bars. A leather sling, or hammock-type seat, is laced to the edges of the bars, and forms a strong base, or ground, for the padded seat—usually made of goatskin stuffed with hair. Figure 2 shows the Walker contract saddle packed for service in the field. Figure 3 shows the parade housing over the Walker saddle. No specimens of this saddle have survived; this drawing was made from descriptions found in the above-mentioned contract and in Hoyt's *Rules for Cavalry*, published in 1816.

Fig. 2. Sergeant, First Regiment, Light Dragoons, in undress uniform, with the Walker contract saddle packed for service in the field, ca. 1812, near side. He is armed with a pair of Model 1805 single-shot muzzle-loading flintlock holster pistols made at Harpers Ferry, West Virginia, and the Starr Model 1812 contract dragoon saber.

Fig. 3. Private, Second Regiment, Light Dragoons, in dress uniform, with the parade housing over his 1812 Walker contract saddle, ca. 1812, near side. He is armed with a pair of Model 1805 single-shot muzzle-loading flintlock pistols made at Harpers Ferry and the Starr Model 1812 contract dragoon saber.

Figure 4 illustrates a second type of saddle used by some troops of United States Light Dragoons during the early years of the nineteenth century. Typical of the hussar-style saddles used almost universally by European armies of this and later periods, it consisted of a wooden three-piece tree with a hammock-type seat (left) and a heavy cloth housing, or shabrack (*schabraque*, right), held in place by its own peculiar style of three-strap surcingle. Figure 5 shows the saddle and equipment as used by field-grade officers.

The saddle shown in Figure 6 is similar to the modern English show saddle with which we are familiar today. It was the standard issue saddle for Light Dragoons near the end of the War of 1812. Equipped with the usual pad for carrying the valise—for the cantle was low compared with the cantle on the saddle in Figure 4, which held the valise clear of the horse's back—this saddle was used in the field with only a folded blanket as a pad under it, but it was equipped with a shabrack for parade and ceremony, as shown in Figures 6 and 7. For the first time small straps at each side of the cantle were added to support extra horseshoes, and a small pouch on each side carried spare horseshoe nails. Later these shoe pouches would enclose both shoe and nails. As was usual until the last few years of the United States horse cavalry, a surcingle was buckled over the saddle to help prevent everything from slipping.

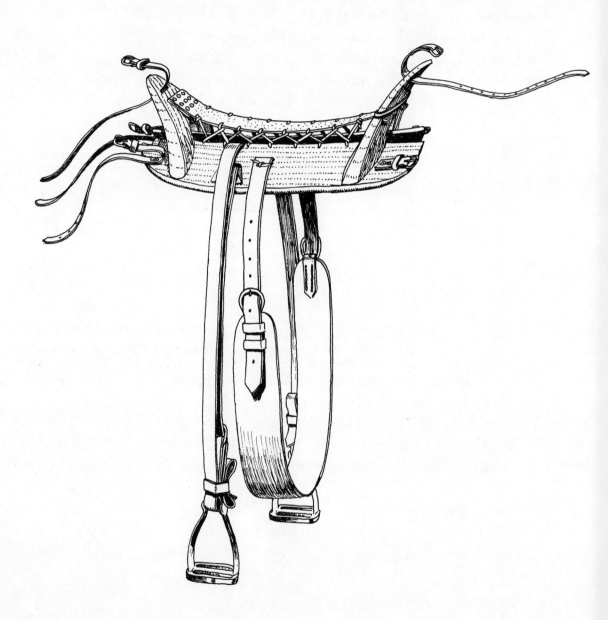

Fig. 4. The hussar-style saddle used by some Light Dragoons, ca. 1812. Opposite: the bare tree. Above: the saddle complete with shabrack and surcingle.

Fig. 5. Lieutenant colonel, First Regiment, Light Dragoons, in full-dress uniform, ca. 1812, near side. His saddle, the hussar style popular with dragoon officers, is equipped with full housing and the distinctive surcingle generally used with the hussar equipment. The bridle is of the same general pattern as the regulation one except that it is more ornate and made of better materials. He is armed with the same weapons as those carried by his junior officers and the enlisted dragoons except that his saber is of the more ornate officer's pattern.

THE WALKER AND OTHER EARLY SADDLES

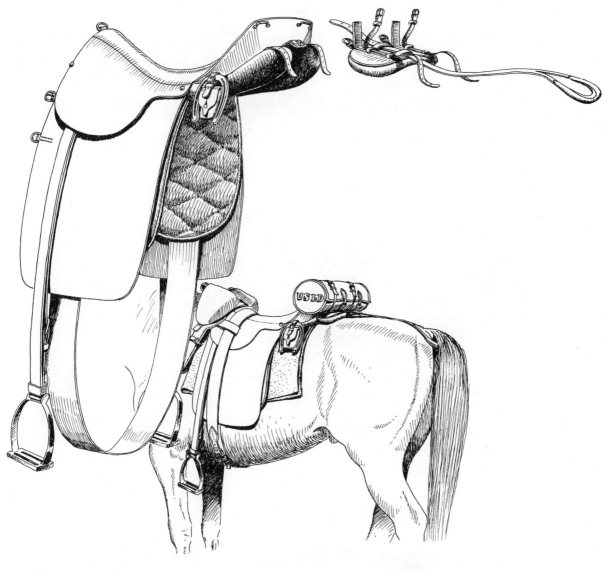

Fig. 6. The issue dragoon saddle and pad for supporting the valise as standardized about 1814 and used until the mounted troops were disbanded in 1816. Above: the saddle as used in the field. Right: the saddle rigged for parade and ceremony.

Fig. 7. Captain, Second Regiment, Light Dragoons, in undress uniform, with one of the issue dragoon saddles packed for service in the field, ca. 1812, near side. The saddle cloth was worn under the saddle for field duty. He is armed with a pair of single-shot muzzle-loading flintlock holster pistols and the officer's model of the Starr Model 1812 contract dragoon saber.

THE WALKER AND OTHER EARLY SADDLES

The mounted arm played little part in the War of 1812. Whether or not the commanders of that period were familiar with the use of cavalry is not known. Perhaps the absence of British cavalry on American soil during the two and one-half years of conflict removed the need for mounted regulars in the eyes of the commanding officers. At any rate, the Light Dragoons saw little if any action during this period. Naturally there was little activity in the design of new types of equipment for a branch of the army on which so little attention was focused. At the end of the war, with a nation pressing for economy, Congress cut the mounted arm to a single regiment of eight troops and in 1816 abolished it completely. Not until 1832 was another mounted force a part of the United States military establishment.

2 DRAGOON SADDLES, 1833-41

Trouble with the mounted Indian tribes of the plains country at last inspired Congress to organize a battalion of six companies of Mounted Rangers in 1832. This most unmilitary-looking force was not furnished with uniforms, arms, horses, or horse equipment by the federal government. The men were obliged by the terms of their enlistment to furnish their own arms, mounts, and equipment. Orders stated that their uniform was to be "the hunting dress of the West."

In the next year, 1833, the Mounted Rangers were disbanded when Congress authorized the organization of the Regiment of Dragoons. Most of the officers and many of the men who had belonged to the Mounted Rangers transferred to the new regiment. The saddles issued to this outfit, which was destined to be the first of the permanent mounted organizations of the United States Army, were almost the same in design as the last issue saddle, which had been standardized in 1814. There were some slight changes, but economy and urgency dictated that the old pattern be changed as little as possible. Thus the horse equipments used by the early dragoon regiments were for all practical purposes about the same as the equipments the Light Dragoons had used about two decades before. No doubt some of the saddles that had been in storage all that time were issued to the first companies of dragoons that were outfitted at Jefferson Barracks, Missouri.

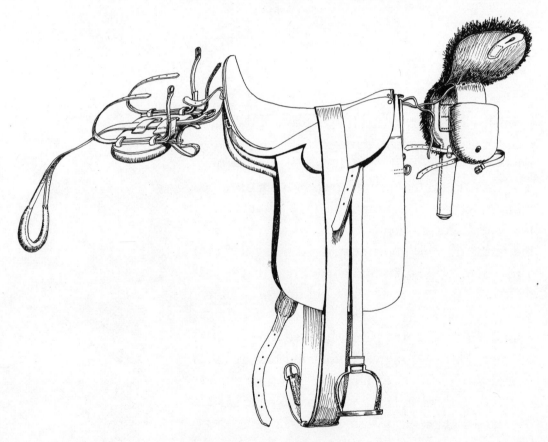

Fig. 8. Above and opposite: The Model 1833 dragoon saddle issued to some dragoon companies until 1844.

DRAGOON SADDLES

Figure 8 shows several views of this very early dragoon saddle. I was fortunate in finding and identifying the only surviving specimen in the Fort Riley (Kansas) Museum at the old cavalry-school site. It had been donated to the collection years before, and to my knowledge it is still displayed in a glass showcase without an identifying label.

Made of black collar leather, the saddle is constructed on a tree similar to the familiar English tree, with a padded underskirt and single billets on each side to fasten to the buckle-ended girth. Rings and

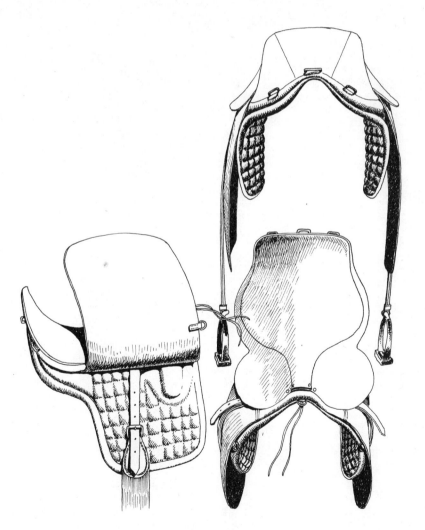

Fig. 9. Sergeant, Regiment of Dragoons, in fatigue uniform, with the 1833 horse equipments, packed for service in the field, ca. 1837, off side. He is armed with the Model 1833 Hall single-shot breech-loading percussion carbine, the Model 1819 S. North single-shot muzzle-loading flintlock pistol, and the Model 1833 dragoon saber.

DRAGOON SADDLES

Fig. 10. Private, Regiment of Dragoons, in fatigue uniform, with the 1833 horse equipments, packed for service in the field, ca. 1837, near side.

staples are attached at both pommel and cantle, to which valise, wallet, holsters, blanket, and cloak were attached for field service. The holster and bag, or pouch, combination, shown in Figure 9, was designed to carry the pistol in the left holster and extra horseshoes, nails, currycomb, and brush in the right-hand pouch, although some holster sets had identical holsters on each side, even though regulations specified that shoes, nails, and grooming tools were to be carried in the right holster. Apparently the holsters were large enough to accommodate this equipment—certainly the single-shot flintlock pistols used then demanded large holsters.

The valise pad was used with this saddle, as shown in Figure 10, since the cantle was not high enough to strap the valise to it directly and still clear the horse's back. A crupper was always used, as was a breast strap, to keep the packed saddle from shifting forward or backward in steep country or at fast gaits.

Newly designed solid-brass-cast stirrups were made for this saddle, and the brass stirrup shown in Figures 11 and 12 was standard for both enlisted men and officers until the McClellan equipments replaced the Grimsley outfit for cavalry in 1859. Some officers used this brass stirrup on their privately purchased saddles as late as the 1880's. I have a pair of these stirrups in my collection.

DRAGOON SADDLES

Fig. 11. Corporal, Regiment of Dragoons, in full-dress uniform, with the half-housing used with the 1833 horse equipments for parade, ca. 1834, off side.

In 1840 or thereabouts a board of cavalry officers was convened at the request of the secretary of war. The board recommended a drastic change in the design of the standard dragoon saddle. The army often works in mysterious fashion, and this incident is certainly hard to understand, for the "new" saddle design was almost identical to the Walker contract saddle of 1812.

Actually, it is doubtful that many of these saddles were made, but the 1841 army regulations included a detailed description of all parts of the saddle, and it is clear that it was intended to replace the former dragoon saddle.

This saddle, shown in Figure 13, was a three-piece wooden saddle, with the tree of bare wood, not covered with rawhide but reinforced with iron plates screwed and riveted to the wooden parts of the tree. Somewhat like that of the old European hussar-type saddle, the seat was formed by a strip of leather stretched between pommel and cantle, with its sides laced to the edges of the bars—a hammock seat. The seat was covered with a pad made of russet sheepskin, lined with canvas, stuffed with curled hair, and quilted. It was fastened to the tree with strings and loops passing over the pommel and cantle and under the girth and girth billet, as shown in Figure 13. Only the cantle was to be bound with brass.

DRAGOON SADDLES

Fig. 12. Lieutenant, regimental staff, Regiment of Dragoons, in full-dress uniform, with the shabrack used by officers over the 1833 dragoon saddle, ca. 1834, near side. He is armed with a pair of Model 1819 S. North single-shot muzzle-loading flintlock pistols and the Model 1833 officer's dragoon saber.

UNITED STATES MILITARY SADDLES

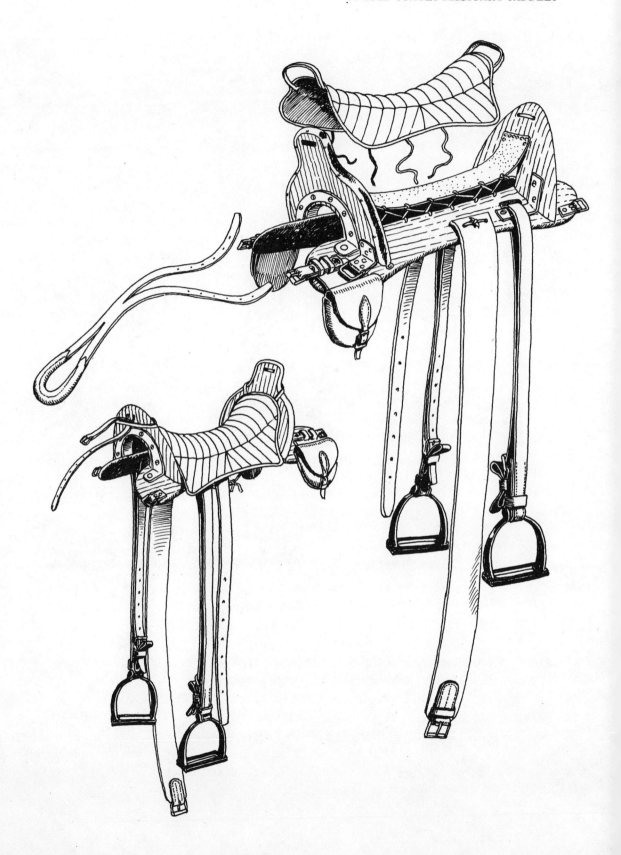

DRAGOON SADDLES

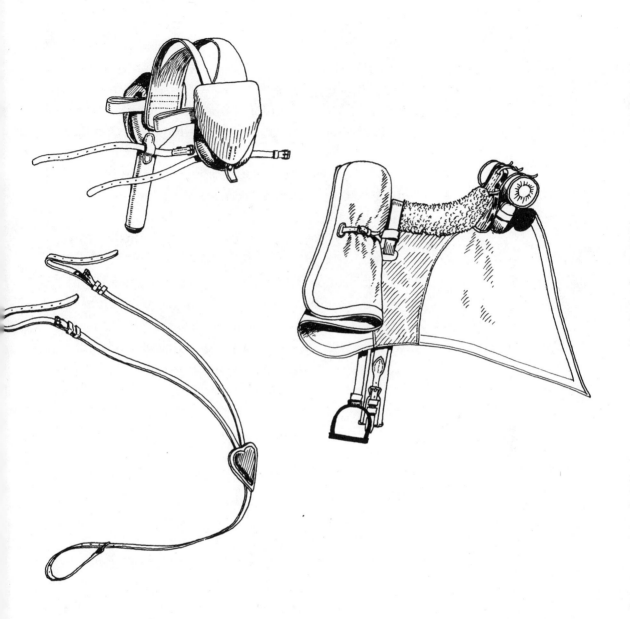

Fig. 13. Opposite and above: The Model 1841 dragoon saddle.

Stirrups were to be of wrought iron, japanned black—I suppost as an economy measure, as must have been the entire saddle, for its construction was certainly cheap as compared with that of the 1814 and 1833 dragoon saddles. Shoe pouches were specified to be mounted at the rear of the sidebars, and, as shown in Figure 14, the holster-pouch combination was to be used with plain leather covers instead of the bearskin that had been specified for the earlier models. Combs and brushes were to be carried in this pouch, as prescribed in the ordnance-manual specifications.

Both crupper and breastplate were specified for use with this saddle. The shabrack, shown at the right of Figure 13, was to be made of dark-blue cloth, lined with canvas, and trimmed with yellow cloth. The seat was to be white sheepskin dressed with the wool on. The shabrack was to be pierced with several slits for passage of the straps of the saddle, and was to be secured to the saddle in front with the two double cloak straps, in the rear with the two double valise straps, and in the center with the surcingle, as shown in Figure 15.

The valise, like the shabrack, was to be made of dark-blue cloth and trimmed with yellow. It was six inches in diameter and twenty-one inches long. It was lined with strong canvas, had the ends stiffened with leather plates, and closed, under the flap, with lacings of strong twine. The outside flap closed with three straps and buckles.

Fig. 14. Private, Second Regiment of Dragoons, in fatigue dress, with the Model 1841 horse equipments, packed for service in the field, ca. 1842, off side. He is armed with the Model 1833 Hall single-shot breech-loading percussion carbine, the 1836 Johnson single-shot muzzle-loading percussion pistol, and the Model 1833 dragoon saber.

Fig. 15. Sergeant, Second Regiment of Dragoons, in dress uniform, with the 1841 horse equipments properly arranged for a ceremonial exhibition, with the shabrack reserved for formal occasions, ca. 1842, off side. He is armed with a modified Model 1833 Hall single-shot breech-loading percussion carbine, the 1805 Harpers Ferry half-stock percussion pistol, and the Model 1833 dragoon saber.

DRAGOON SADDLES

No specimen of this saddle exists, and I doubt that any but ordnance patterns were made. During this period Congress was taking extraordinary economic precautions, and every possible means was employed to cut the expenses of maintaining a national army. Thus it was that inferior arms and equipment held back progress in the young army and made the keeping of domestic peace and tranquillity doubly hard for the handful of mounted regulars scattered over the vast frontier.

3 THE RINGGOLD AND GRIMSLEY SADDLES, 1844-47

A new saddle was adopted by the United States Army in 1844, even though the two dragoon regiments had for several years been engaged in comparatively peaceful operations (the Second Regiment of Dragoons had been formed in 1836 to participate in the Seminole Indian Wars in Florida). Army Major Samuel Ringgold, a regular artillery officer, designed an improved dragoon saddle that was suitable for both dragoons and field artillerymen and was successful in having a board of officers recommend it for adoption. Many of these saddles were made at the Quartermaster Depot in Philadelphia, as well as by private contractors.

Figure 16 shows three views of the Ringgold saddle, as drawn from the specimen in the West Point Museum. The tree of this saddle is made of ash, and consists of six pieces: the two sidebars, the pommel in two pieces forming an arch over the withers of the horse, and the cantle in two pieces forming an arch over the backbone. The two halves of both pommel and cantle are joined so that the grain is at right angles, and both pommel and cantle are reinforced with iron straps riveted to the wood. Pommel and cantle are fastened to the sidebars with iron pins riveted through the wood of the bars and the sheet-iron angles to which both pommel and cantle are also riveted. Brass moldings on pommel and cantle protect the end grain of the wood from the weather. The sidebars in front of the pommel and behind the cantle are covered with sheet iron, through which staples and ring staples are riveted.

An iron plate, U-shaped, is attached to each sidebar to hold the girth billets, and an iron plate fastened to the bars just above the stirrup strap mortise strengthens this part of the tree and affords a strong brace for the strain placed on the stirrup leathers and that part of the bars.

The seat is covered with strong webbing firmly stretched and attached to both pommel and cantle. The webbing is in turn covered with rawhide. Both are attached to members of the tree with copper nails. Both inner and outer skirts are fastened to the sidebars with copper nails. The outer skirt has a slit for the passage of the surcingle, as shown in the center drawing. This arrangement was designed to prevent the chafing of the rider's legs. The seat is sheepskin with a canvas lining, stuffed with curled hair and quilted. It is permanently fastened to the tree with brass saddler's tacks.

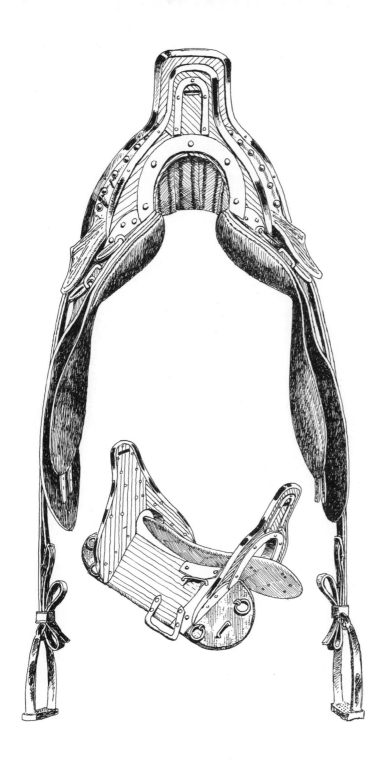

Fig. 16. Above and on page 36: The Ringgold dragoon saddle adopted by the army in 1844.

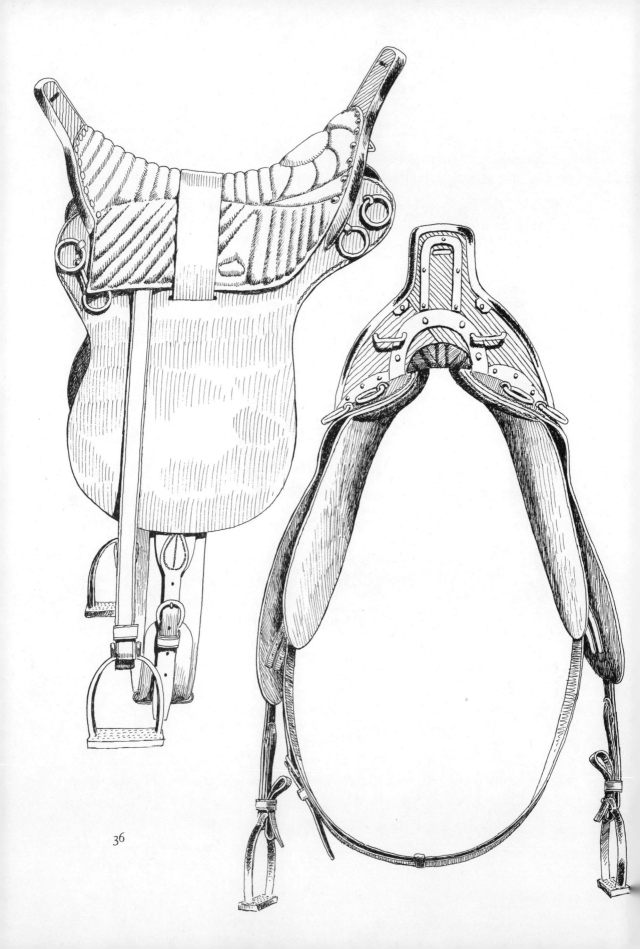

THE RINGGOLD AND GRIMSLEY SADDLES

The stirrups on the saddle in Figure 16 are those intended for artillery, being brass, as were those for dragoons, but with the treads roughened instead of perforated like the 1833 stirrups, shown in Figure 8. This saddle was certainly an improvement over the Ordnance Model 1841, shown in Figure 13, and a big departure from the 1833 English-type saddle, shown in Figure 8. Major Ringgold, describing his saddle in a letter to the quartermaster general before it was adopted by the army, wrote:

> In constructing a military saddle and arranging the equipments there are certain indispensable conditions that must be fulfilled. Among the rest may be enumerated:
>
> 1. Protection to the horse from injury by a proper formation of the saddle.
> 2. Transportation of the effects of the soldier without embarrassment to man or horse in travelling, maneuvering, and the use of weapons.
> 3. Durability, strength, and a view to proper economy.
> 4. Fitness for campaign and war.
> 5. As much as possible the ease and comfort consistent with a correct military seat.
>
> It is confidently believed that the saddle I have the honor to offer for inspection fulfills these conditions. The cantle is no higher than necessary to sling a valise clear of the loins. The pommel is no higher than to raise the arch over the withers and carry the holster and cloak free of pressure.

Figures 17 and 18 show the saddle with full pack for field service.

Later the same year, on October 7, 1844, Patent No. 3779 was granted to Major Ringgold, after his saddle had been adopted as the standard saddle for the mounted services of the United States Army. But the Ringgold saddle was not the most popular piece of equipment with officers or enlisted men of either the artillery or the dragoon branches. This saddle was the most widely used throughout the Mexican War, but many officers bought other kinds of saddles, including stock-type saddles with horns, eagle-head-pommel saddles, and a new dragoon saddle on an especially strong tree made by Thornton Grimsley, of St. Louis. Many glowing letters of praise were sent to the quartermaster general by both artillery and dragoon officers who had used the Grimsley saddle during the war—when Grimsley let it be known that he was pressing for adoption of his saddle in 1847.

A year earlier, on September 1, 1846, Colonel Aeneas Mackay, assistant quartermaster at the St. Louis Quartermaster Depot, had written to Major General Thomas S. Jesup, the quartermaster general, that Grimsley was then finishing the saddle and related equipment for one company of First Dragoons who had been recruiting for some time, as well as the saddles and other horse equipments for the Regiment of Mounted Rifles, which had just been authorized by Congress. It is evident that Grimsley had supplied a large number of saddles to the army a year before his design was officially adopted.

In the same letter Colonel Mackay mentioned that Colonel Stephen Watts Kearny, in command of the Army of the West in 1846, had taken a number of Grimsley saddles with him on his march to Santa Fe and California, and "he never made any objection to

Fig. 17. Private, Second Dragoons, with the regulation Ringgold horse equipments and full pack for field service, ca. 1846, near side. He is armed with the Model 1842 single-shot muzzle-loading percussion pistol, the Model 1843 Hall single-shot breech-loading percussion carbine, and the Model 1840 dragoon saber.

Fig. 18. Private, Second Dragoons, with the regulation Ringgold horse equipments, ca. 1846, off side.

THE RINGGOLD AND GRIMSLEY SADDLES

them and the popularity of them has become so great that I have been obliged to prohibit the private sale of them to the officers and others, to enable Mr. Grimsley to complete on time the number required by the dragoons."

Figure 19 shows the Grimsley dragoon saddle in two views with no field equipment and one view packed with valise, shoe and nail pouches, crupper, breastplate, holsters, and cloak, and with the same type of carbine bucket that had been in use since the Light Dragoons used it in the American Revolution more than seventy years before.

This is the basic saddle design that Colonel Kearny submitted to the quartermaster general in 1844. Of the tree Grimsley had the following to say in a letter to the quartermaster general after the board had recommended his saddle for adoption:

> The tree introduced by General Kearny was a complete compilation of the French and Spanish and made by myself. The one now submitted is my improved tree, and the Board specially referred to my improvements.

The board responsible for the adoption of the Grimsley was filled with distinguished names, including Brigadier General Kearny, formerly colonel of the First Dragoons; Major Thomas Swords of the Quartermaster Department; Major Phillip St. George Cooke of the Second Dragoons and later in charge of the Union Cavalry during the American Civil War (and also the father-in-law of that great Confederate cavalry leader J. E. B. Stuart); Captain and Brevet Lieutenant Colonel Charles A. May, Second Dragoons, the hero of the charge against the Mexican artillery at Reseca de Palma during the Mexican War; and Captain H. S. Turner of the First

UNITED STATES MILITARY SADDLES

Dragoons. All these officers, who had seen active service with the dragoon regiments, had personal experience with the Grimsley and were enthusiastic about its merits.

The Grimsley tree was the first United States military saddle to be encased in wet rawhide; thus it was

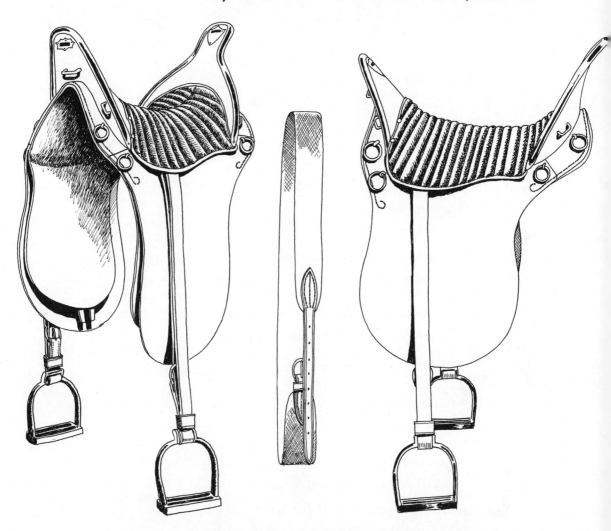

Fig. 19. Above and opposite: the Grimsley dragoon saddle adopted by the army in 1847. Above: the Grimsley saddle and its equipments packed for field duty and the Grimsley tree.

THE RINGGOLD AND GRIMSLEY SADDLES

an extremely strong tree, much less susceptible to damage from the rigors of campaign than any of the former all-wood trees. It was lighter than the Ringgold, for the rawhide cover removed the necessity for the heavy reinforcing iron bands that had been used on the Ringgold.

In its report to the adjutant general's office the board said:

It is the unanimous opinion of the Board that the best pattern of saddle for cavalry service is the one submitted by Colonel (now Brig. Gen.) S. W. Kearny to the Quartermaster General in 1844, with the changes made by Mr. Grimsley since that time. A complete model of the saddle, as modified, was submitted to the Board by Mr. Grimsley, and they recommend its adoption, as it is, for Cavalry Corps. Combining strength, durability, peculiar fitness to the horse's back and convenience for military fixtures, this pattern more than any other yet furnished for Dragoon service, gives an erect posture, and easy seat to the rider, at the same time that little or no injury is done to the horse's back on the longest marches. Some of the members of the Board have had the fairest opportunity of testing the merits of this saddle, having used it on marches of more than 2,000 miles in extent, and the result has been in every instance to confirm their belief in the superiority of this saddle over any other which has come under their observation. In outward appearances this saddle resembles the French Hussar saddle more than any other with which the Board is familiar; it combines all the conveniences of the French saddle for attaching military and cavalry appurtenances, with the indispensable qualities requisite in a service tree. To prevent injury to the horse's back the "side-bars" are so formed as to *fit* the back bearing equally throughout their whole extent; and the forks of the high pommel and cantle are, in every case, and under all the circumstances of reduced flesh, raised above the withers and backbone of the horse.

THE RINGGOLD AND GRIMSLEY SADDLES

Fig. 20. Corporal, Regiment of Mounted Rifles, with full field pack on the new Grimsley dragoon saddle with which the regiment was equipped, ca. 1846, near side. He is armed with the single-shot muzzle-loading Model 1841 rifle (the "Mississippi rifle"), the Model 1836 single-shot muzzle-loading flintlock pistol, and the Model 1840 heavy dragoon saber.

The Grimsley tree is covered with black collar leather, and has a quilted seat. Two sets of skirts are attached to the tree: the underskirts to protect the sides of the horse from the girth buckles and the outer skirts to protect the rider's clothing from horse sweat. Figure 20 illustrates the Grimsley dragoon saddle with full field pack.

The peaks of both pommel and cantle are mortised for cloak and valise straps, and the mortises are enclosed on the outside with eight-pointed brass escutcheons and on the inside with brass shields. Staples are set in both pommel and cantle for cloak and valise straps and for securing holsters to the saddle. Rings are held to the ends of the sidebars with staples riveted to the bars. The stirrups are cast brass and are the same pattern as those furnished to the Regiment of Dragoons in 1833 and 1834, as shown in Figure 21.

Girth and surcingle are made of indigo-blue worsted webbing with leather buckle and billet ends. The girth is three and one-half inches wide and three feet nine inches long. The surcingle is the same width as the girth, and the webbing part is five feet long, and the leather end of the billet is two feet long.

The Grimsley saddle was the regulation saddle for all mounted personnel until the McClellan saddle was adopted in 1859. Many officers continued to use their Grimsley saddles throughout the four years of the Civil War, including Generals Ulysses S. Grant and William Tecumseh Sherman.

Fig. 21. The 1847 Grimsley dragoon officer horse equipments, near side. The officer wears the uniform that was regulation from 1833 to 1851.

4 THE CAMPBELL, HOPE (TEXAS), AND JONES SADDLES, 1855-58

In the middle of the nineteenth century the Plains Indians realized that their hunting grounds were being taken over by the encroaching whites. The whole frontier erupted in sporadic raids by war parties. The three mounted regiments were scattered from Texas to Oregon, and Congress finally realized that the army at its then-existing strength could not cope with an "Indian problem" now grown into a serious threat. In March, 1855, Congress authorized two new bodies of mounted troops. For some reason not clear to military historians the two new regiments were designated the First and Second Regiments of Cavalry, instead of Dragoons or Mounted Rifles, as the older mounted regiments were termed. Jefferson Davis, later to become the president of the Confederacy, was secretary of war and was afterward accused of staffing the new cavalry regiments with favored southern officers, who joined the Confederacy when the Civil War split the nation.

The formation of the two new regiments gave the quartermaster and ordnance departments a good excuse to field-test new items of uniform, arms, and horse equipments. On August 15, 1855, General Order No. 13 was issued by the Adjutant General's Office, in which specific instructions were given for equipping the First and Second Regiments of Cavalry, authorized only five and a half months earlier.

Four squadrons (one squadron consisted of two companies) of each of the two new regiments were to be furnished with the Grimsley equipments used by the older dragoon and mounted rifle regiments, as shown in Figure 19. The remaining, or fifth, squadron of each regiment was to be supplied with the experimental Campbell saddle, modified from the original form as shown in the patent drawings to conform to specific Ordnance Department specifications. Just a few months earlier, on May 21, 1855, General Order No. 5 had transferred the responsibility of horse-equipments procurement and supply from the Quartermaster Department to the Ordnance Department, where it was to remain for many years.

Figure 22 shows the Campbell saddle as it was patented by Daniel Campbell in 1855. His first patent covered only the tree with its self-adjusting sidebars. The French cavalry already had a saddle with this feature, which had been designed by one Captain Nugent. While the adjustment was accomplished by a slightly different means, the end result was the same: no matter how much the flesh of a cavalry horse fell off during the exertions of a campaign, the sidebars would automatically adjust so that the fit on the horse's back would always be perfect, and sore backs from poorly fitting saddles would, at least in theory, become a thing of the past. The one serious flaw, however, was the inevitable weakening of the tree—and that was the case with the Campbell saddle, for it did not survive this extensive field test.

The drawings of the tree at the left of Figure 22, copied from the July 10 patent, show how the spring-steel arches connecting both pommel and cantle to the sidebars made possible the automatic adjustment of the angles of the sidebars.

The second patent, issued on December 4 of the same year, covered Campbell's improved dragoon saddle, as shown in the two views in the center of Figure 22. This patent covered not only the saddle itself but also the design and attachment of holsters and valise. Both these appurtenances were attached primarily to the sidebars so that they would not impair the adjustment of the sidebars to the conformation of the individual horse. The valise and also the holsters were each made in two pieces connected with flexible leather. An added and novel feature of the holsters was the cover, which swiveled forward on a metal pivot to allow the cavalryman to remove the pistol from its holster.

The original saddle had pommel and cantle shaped much like those of the Grimsley—both peaked and fairly high and both well sloped. The stirrups on the original design were cast brass, like those used since 1833. But the army required that the test saddles be made with both pommel and cantle more nearly vertical and that the peak of the cantle be removed so that the top of the cantle would be

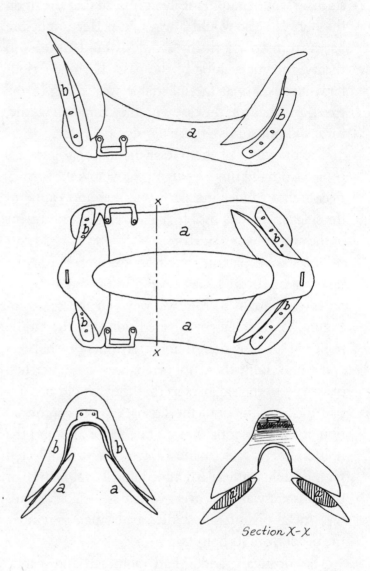

Fig. 22. The self-adjusting saddle designed by Daniel Campbell and patented in 1855. Above: diagrams made from Campbell's original patent. Opposite, left: the saddle in its original form. Opposite, right: the improved Campbell saddle as altered for field testing by cavalry units in 1855.

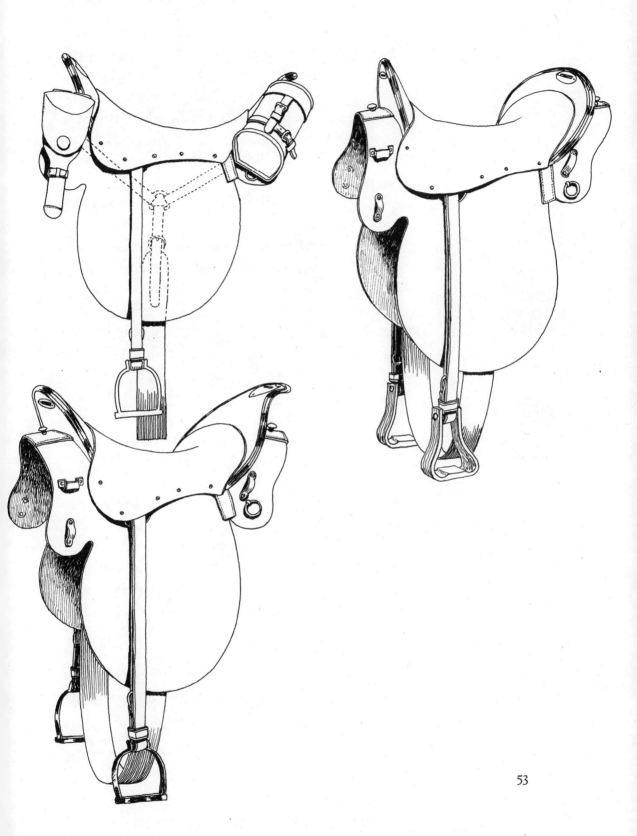

rounded, like the sides. The brass stirrups were replaced with open wooden stirrups—the same kind, presumably, that were used on the McClellan adopted by the army four years later. The drawing at the right of Figure 22 shows the Campbell saddle as it was issued to the First and Second Cavalry. Figures 23 and 24 show the saddle with full pack and Campbell horse equipments.

That the Campbell saddle did not prove satisfactory in the extensive field tests it received to warrant adoption for the mounted service must be a foregone conclusion, since it was replaced within a year by the standard Grimsley. No doubt reports from cavalry officers, if they could be found, would reveal that there was excessive breakage of the trees under the rigorous use the First and Second Cavalry gave them.

The Campbell was not the only saddle given extensive field tests during this period. While the new First and Second Cavalry were riding the Campbell saddle in Texas and Kansas during the winter of 1855–56, the officers of these regiments, who were required to purchase their own saddles, seem to have been more impressed by the saddles used by the ranchers and frontiersmen than with those favored by the military. The Hope saddle, also known as the Texas saddle, was a favorite. Joseph E. Johnston, then commanding the Second Cavalry, wrote the secretary of war in 1856 that all the officers who could obtain Hope saddles did so.

Other officers on both sides during the Civil War used variations of the Hope saddle. Figure 25 was drawn from Johnston's personal saddle, now in the

Fig. 23. Private, First Cavalry, with full pack and the Campbell horse equipments issued to one squadron of each of the two new cavalry regiments for trial in the field, ca. 1855, near side. He is armed with the saber and with the Model 1855 Springfield pistol-carbine with detachable stock. Stock and pistol are carried detached in the special pommel holsters.

Fig. 24. First sergeant, Second Cavalry, with full pack and the Campbell horse equipments issued to one squadron of each of the two new cavalry regiments for trial in the field, ca. 1855, off side. He is armed with the Colt Navy revolver, the saber, and the Model 1854 muzzle-loading percussion carbine.

THE CAMPBELL, HOPE, AND JONES

collection of the Confederate Museum in Richmond, Virginia. Figure 26 shows the saddle belonging to Major General Hugh Judson Kilpatrick, a Union cavalry leader. The trees in Figures 25 and 26 are identical, the difference in the two saddles being the shape of the skirts and the decorative stamping.

In 1857 the chief of ordnance reported that his department had ordered manufactured and put into service in the field 170 sets "of the pattern known as Hope's saddle and much used in Texas.... The use of these [the Campbell, Hope, and Jones models] and the Grimsley equipment in actual service will afford the means of comparing their actual merits."

The only available records (Ordnance Department correspondence) show that the Hope saddle was issued to two companies of the Second Cavalry and that the equipment was procured from Rice & Childress, a San Antonio, Texas, firm (letter dated February 10, 1857, from Major A. K. Craig, Ordnance Department, Washington, D.C., to Messrs. Rice and Childress, San Antonio, Texas, records of the Office of Chief of Ordnance, Record Group 156, National Archives).

Figure 27 shows the Hope saddle as it must have been issued to the Second Cavalry companies, drawn from a specimen in the Fort Sill, Oklahoma, Museum. The tree of the Fort Sill saddle and the trees used by Kilpatrick (in the National Museum) and Johnston (in the Confederate Museum) are identical. The issue saddle has no skirts, but is covered with black collar leather now turned brown from age. All three specimens have trees with slots through which the stirrup leathers are hung, and all three have essentially the same dimensions.

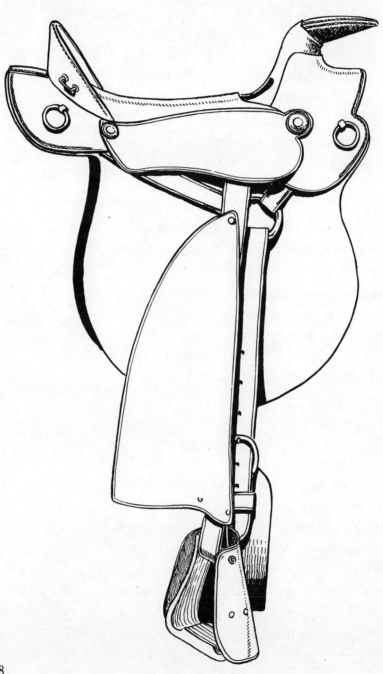

Fig. 25. A privately purchased Hope saddle owned by Confederate General Joseph E. Johnston, now in the collection of the Confederate Museum, Richmond, Virginia.

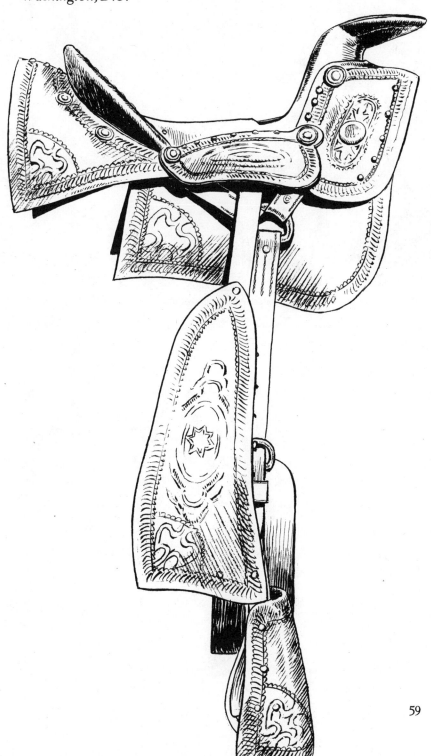

Fig. 26. The Hope saddle owned by the Union cavalry leader General Hugh Judson Kilpatrick, now in the Smithsonian Institution, Washington, D.C.

Fig. 27. The Hope saddle as issued to two companies of the Second Cavalry in 1856 or 1857, drawn from the specimen in the Fort Sill Museum collection.

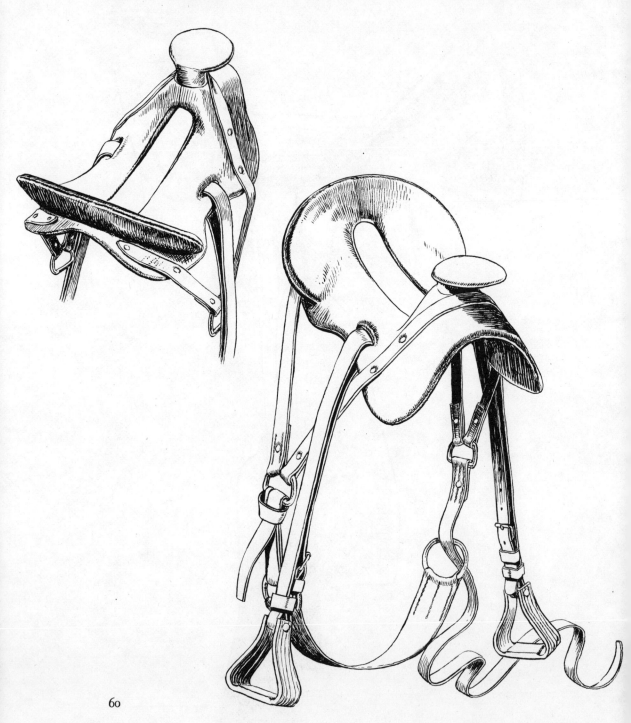

THE CAMPBELL, HOPE, AND JONES

The Jones saddle mentioned above by the chief of ordnance was the invention of First Lieutenant William E. Jones of the Regiment of Mounted Rifles. Figure 28 shows his patented adjustable steel frame, as shown on Patent No. 11,068, granted on June 13, 1854. No. specimen of this saddle seems to exist, and its completed form must remain speculation, although it seems logical that it must have been similar in appearance to the Grimsley, then the regulation saddle.

Fig. 28. The Jones steel-frame adjustable saddletree from the patent granted to Lieutenant William E. Jones, of the Regiment of Mounted Rifles, on June 13, 1854.

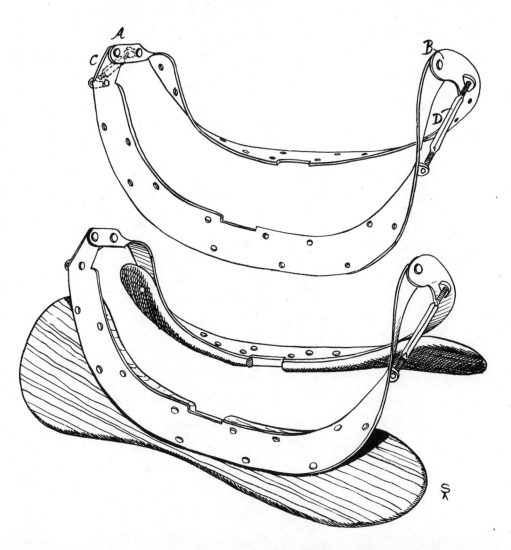

Records in the National Archives show that the quartermaster general approved the purchase of three hundred Jones saddles, which were issued in 1857 or 1858. Since no more were purchased, it is apparent that the technology of the times was insufficient to produce spring steel strong enough to stand up under hard field use and that the saddle proved unsatisfactory for general adoption.

5 THE McCLELLAN SADDLE, 1859

In 1855, during this period of testing, Captain George B. McClellan of the First Cavalry was sent to Europe as a member of a commission of American officers to study the armies of Europe and their equipment—actively engaged at this period in the Crimean War. After McClellan's return his report was published in book form as a Senate document in 1857. It was filled with descriptions and recommendations for changes. The next year he had several saddles made that incorporated his ideas for the ideal military saddle, and in 1859 a special board of officers recommended that the Grimsley saddle should be replaced by McClellan's new design.

Captain McClellan always asserted that his saddle was the result of intensive study of European saddles, and more often than not military writers quote him as claiming that he used the best features of the Hungarian and Mexican trees to perfect his design. It is strange to me that no military writer has seen through this bit of subterfuge on McClellan's part and revealed the true source of the McClellan design.

Close examination of the illustrations of the Grimsley saddle, the Campbell saddle, the Hope saddle, and the first McClellan saddle reveals that the McClellan saddle incorporates the major features of the Grimsley, the Campbell, and the Hope. McClellan's tree is shaped almost exactly like the Campbell tree as modified for issue to the First and the Second Cavalry for field trial and the Hope tree. The rigging —two quarter straps joined at a D-ring hung below the approximate center of the tree and buckling to a girth on each side—is identical to the rigging on the Hope and the Campbell. The tree itself is covered with wet rawhide, just as the Grimsley and Hope are, and the staples, rings, and pommel and cantle mortises are little different—even to the brass escutcheons at the mortises. Even the stirrups on the McClellan are the same as those on the modified Campbell, with the addition of leather hoods.

Of course, there are obvious differences between the McClellan and the other saddles—the absence of a leather covering on the seat and bars; a single, different-shaped skirt *under* the quarter straps—an obvious measure of economy, eliminating the necessity for additional leather for seat and an underskirt, since the McClellan's single skirt protects the horse from the rigging and the rider's legs from horse sweat.

The girth is different, too, but it still buckles at both ends and is made from the same indigo-blue worsted webbing, as is the surcingle, whose design remained the same as the one used with the Grimsley saddle.

THE MCCLELLAN

There is also a difference in general appearance, but from a design standpoint the McClellan came very close to the modified Campbell and the Hope—a far cry from the Hungarian hussar-saddle–Mexican-tree combination claimed by Captain McClellan. The Grimsley was made like the Mexican, or Texas, tree in that it was strengthened with a similar covering of rawhide sewed on wet, but that is where even the Grimsley's resemblance to the Mexican tree ended. There is no doubt in my mind that McClellan borrowed his saddle design much closer to home than he claimed. Neither Campbell nor Grimsley seems to have made a big outcry of objection as far as records in the National Archives reveal, but perhaps that is because both men died soon after McClellan's saddle was adopted.

Figure 29 illustrates the father of the modern McClellan. This saddle lasted, virtually unchanged, for almost one hundred years of use by the United States Cavalry. In those many years it lost its skirts but finally regained them in 1928. The tree was fully covered with leather just ten years after its adoption by the army, and stirrup hoods were removed and put back several times before horse cavalry became a thing of the past, and the rigging was changed slightly a number of times.

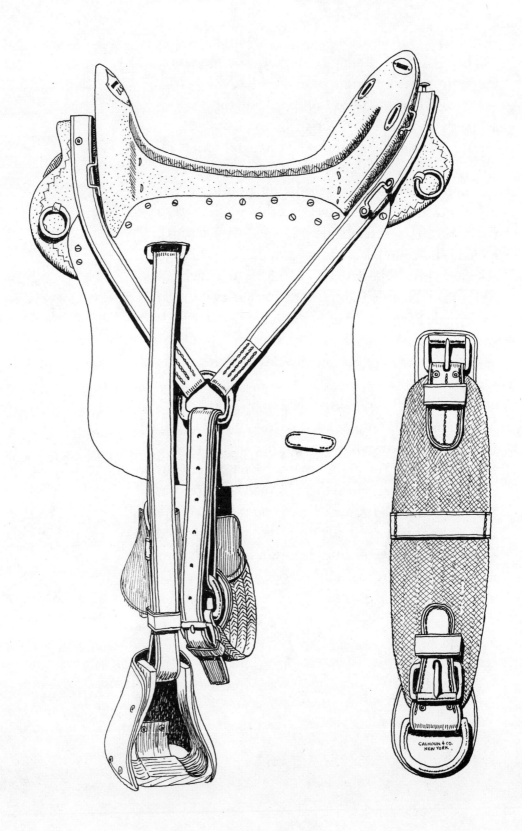

THE MCCLELLAN

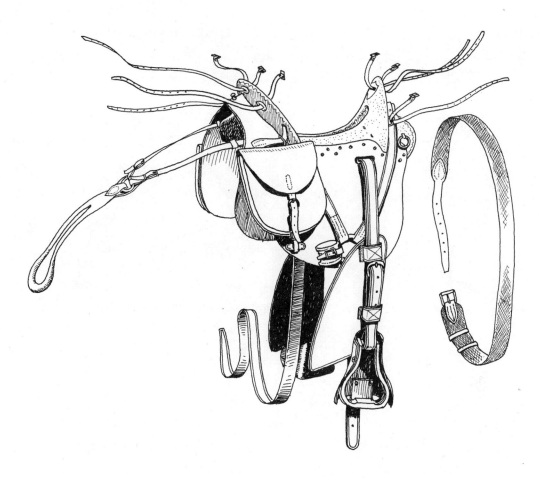

Fig. 29. Opposite: The Model 1859 saddle for United States mounted forces—the first McClellan. Above: the McClellan saddle with cavalry equipments used from 1859 through the mid-1870's. The saddle included an exposed rawhide tree, and all leather was dyed black.

Fig. 30. Sergeant, First Cavalry, with full pack and Model 1859 McClellan horse equipment, ca. 1864, off side.

THE MCCLELLAN

That the McClellan remained the standard saddle for all United States mounted forces for so many years certainly was not because better saddle designs for cavalry were not forthcoming periodically. Its long tenure was the result of economics. There were so many thousands on hand after the Civil War that each time a change was made the frugal powers that dictated armed forces expenditures never could justify replacing the vast stocks of McClellans in the nation's armories with the many improved designs that were given field trials every few years.

Figure 29 shows the exposed rawhide tree of the Model 1859 McClellan. The concern for preserving horseflesh rather than the skin of the cavalryman's posterior is clearly evident: all seams were carefully placed on the rawhide tree cover so that they could not possibly chafe a horse's back or even a saddle blanket. But the seams at the edge of the split on the seat and around the edge of the cantle were especially affected by continuous wettings by rain and exposure to direct sun rays, and every saddle that was used much in the field soon had split seams in these areas, which proved to be hard on the trooper if not on his horse. Otherwise, the first McClellan was a good wartime saddle and served both northern and southern armies well.

The drawing at the right of Figure 29 shows the saddlebags, crupper, coat straps, sweat leathers, and surcingle that were standard McClellan equipments. The girth for all issue saddles was made of indigo-blue worsted webbing with a buckle at each end. It was four and one-half inches wide and about fourteen inches long over all. Some officers' saddles had girths of the same pattern and dimensions made of heavy harness leather. The Model 1859 McClellan in my collection has such a leather girth. Figures 30 and 31 show the McClellan horse equipment.

Figure 32 shows several of the many variations and modifications of the McClellan that were found among officers' saddles used throughout the four years of the Civil War and on into the 1870's wherever officers were on horseback. The saddle at the left is made on the standard McClellan tree, but has a full leather covering and double skirts with the girth billets and D rings coming through slits in the underskirt. Both cantle and pommel are brass-bound, and the girth is leather. The saddle at the lower center has a full leather-covered tree with a quilted seat, and the pommel and cantle are brass-bound. It too has a leather girth of standard size and pattern.

The McClellan at the top right of Figure 32, drawn from Major General John Sedgwick's saddle in the West Point Museum, shows it fitted with a major general's shabrack, holsters, breastplate, crupper, and valise. The saddle itself is almost identical to the one shown here with the quilted seat. The girth on this specimen is also leather. There were many other variations, most of them much like the saddles shown here.

Fig. 31. Corporal, First Cavalry, with full pack and Model 1859 McClellan horse equipment, ca. 1864, near side.

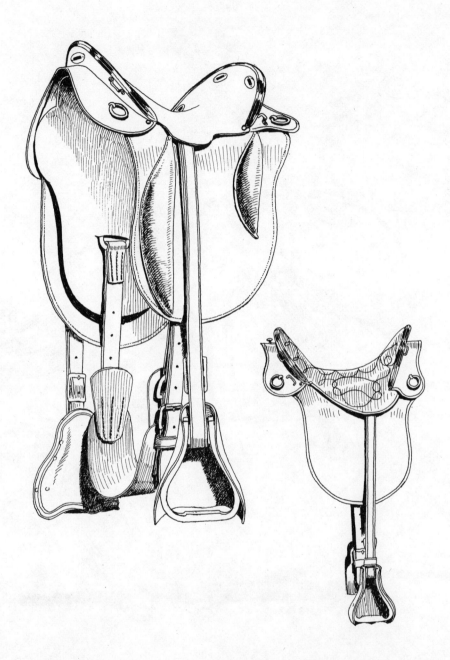

Fig. 32. Above and opposite: modified McClellan saddles used by commissioned officers, 1860–75.

THE MCCLELLAN

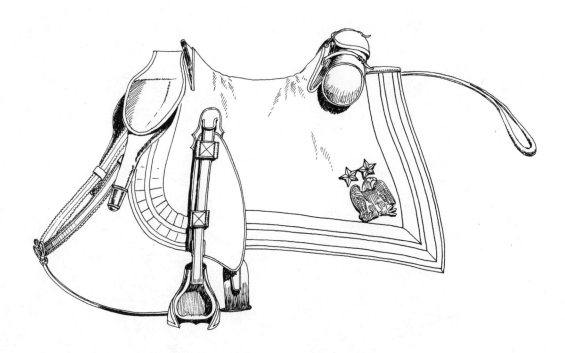

Confederate-made McClellans were much in evidence among officers in the southern armies. The leather on these saddles was almost always russet-colored, since the means to dye leather black were scarce. Some Confederate McClellans have been described as having skirts made of several layers of linen cloth stitched together, for leather was not in abundant supply during the latter part of the war. Most Confederate cavalry McClellans were captured from a well-equipped adversary.

6 THE JENIFER SADDLE, 1860

The McClellan saddle was not the only saddle ridden by officers of both sides during the Civil War. Many officers rode on military-type saddles built on Texas trees with horns—the Hope saddle.

Figure 33 shows the Jenifer saddle, which was extremely popular with both northern and southern officers. While it was never officially adopted by the northern army, it was submitted for trial and consideration in 1860. Designed by Walter H. Jenifer, of Baltimore, it was in fairly common use among mounted officers a year or more before a patent was granted to Jenifer in mid-1860. The Confederates thought enough of the Jenifer pattern to make substantial numbers of the saddle during the war, according to the report of the Confederate quartermaster general. But the same report admitted that the Jenifer caused more sore backs than did the Confederate-made McClellans.

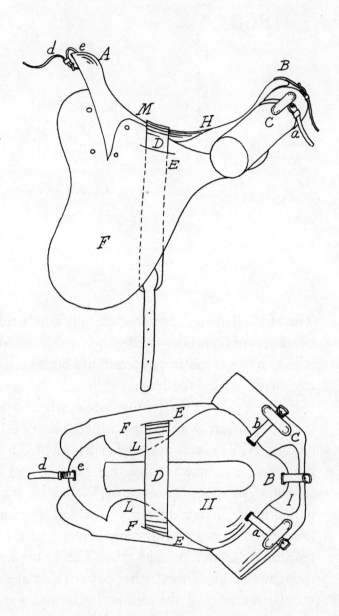

THE JENIFER

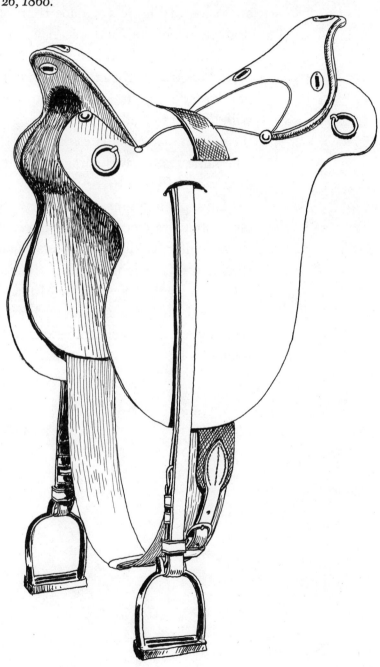

Fig. 33. The Jenifer cavalry saddle of 1860. This saddle was used by some Union officers and Confederate cavalrymen during the Civil War. The tree design at the left (opposite) was drawn from the patent issued to Walter H. Jenifer on June 26, 1860.

The patent issued to Jenifer was for "a new and improved military saddle" and included claims for innovations both in the saddle and in the valise, which was made to fit the curve of the cantle. The drawings at the left of Figure 33, made from Jenifer's patent drawings, show top and side views. The drawing at the right was made from a specimen in the Confederate Museum in Richmond, Virginia. One feature of this saddle was the slit in the skirt near its juncture with the seat that allowed the surcingle to pass under the skirt so as to eliminate chafing of the rider's legs. The same feature was part of the Ringgold saddle, as shown in Figure 16.

Many photographs by Mathew B. Brady and other photographers of the Civil War period clearly show Jenifer saddles on officers' mounts. Another specimen of the Jenifer can be found in the study collection in the West Point Museum. Both metal and wood stirrups seem to have been in equal favor with owners of Jenifer saddles, according to existing photographs.

7 MODIFICATIONS OF THE McCLELLAN SADDLE AND THE WHITMAN SADDLE, 1868-1904

After the close of the Civil War the Union cavalry regiments were widely dispersed throughout the United States (at the beginning of the war all regiments of Union mounted corps had been designated "cavalry," eliminating the nomenclature "dragoons" and "mounted rifles"). In the same year, 1861, a sixth regiment of mounted soldiers, the Sixth Cavalry, was authorized by Congress. In 1866 four additional regiments of cavalry were authorized and recruited to take charge of the growing Indian problem in the West. Two of these regiments, the Seventh and the Eighth, were all-white; the Ninth and Tenth Regiments were recruited from all-Negro sources for enlisted and noncommissioned personnel; all commissioned officers of these two regiments were white.

Until 1868 the issue saddle for all regiments was the Model 1859 McClellan. The saddles were left over from the vast stores made for the war and were unchanged in any respect from the thousands of saddles that had carried the cavalrymen through four years of fighting.

In 1868 the Ordnance Department recognized that the war saddles were lasting only a few months in the climatic extremes prevalent in the West, and a series of modifications was begun that attempted to remedy the shortcomings of the exposed rawhide-seat Model 1859 McClellan.

An ordnance board, convened in January and February, 1868, published its recommendations in Ordnance Memorandum No. 9. The board recommended that no change be made in the standard (wartime) cavalry equipment until the views of the officers of the cavalry arm of the service had been obtained on the subject. Monthly reports from company commanders were requested on the following points: "Whether the stirrup-hood or sweat leathers, or both, may be dispensed with; whether the saddle flaps [skirts] may not be removed or reduced in size; and whether breast straps and cruppers should be used."

Because of the large quantity of cavalry equipments on hand, including saddles, the board recommended that no further repairs be made to the old ones until the requested information had been received and that the new saddles then in stores, before being issued, should have both cantles and pommels covered with brass molding according to an approved pattern. The upper left-hand drawing of Figure 34 shows these brass moldings in place on the Model 1859 saddle, held in place with small brass nails and screws at the terminal points of each molding.

MODIFICATIONS OF THE MCCLELLAN AND WHITMAN

During 1869 many company commanders in the field requested supplies of "fair" (russet) leather with which to cover the trees of their saddles, using their own company saddlers for this task. The brass moldings on the saddles requisitioned to replace unserviceable company saddles had not been satisfactory, and seams along the seats still split open and chafed human skin raw. The War Department approved such requests, and hundreds of saddles had skirts removed temporarily while the trees were covered; then the skirts were fastened back in place with the same small brass screws. A saddle with the tree covered with russet leather is shown at the lower left in Figure 34.

In the next year, 1870, an Ordnance Department memorandum recommended that one hundred saddles with leather-covered trees be issued to each regiment for trial, since insufficient information had been received from officers in the field to make a general decision about covering all saddles with leather. Two years later, on June 29, 1872, General Order No. 60 ordered all saddletrees covered with black collar leather and the addition of a square leather safe where the quarter straps were stitched to the D ring.

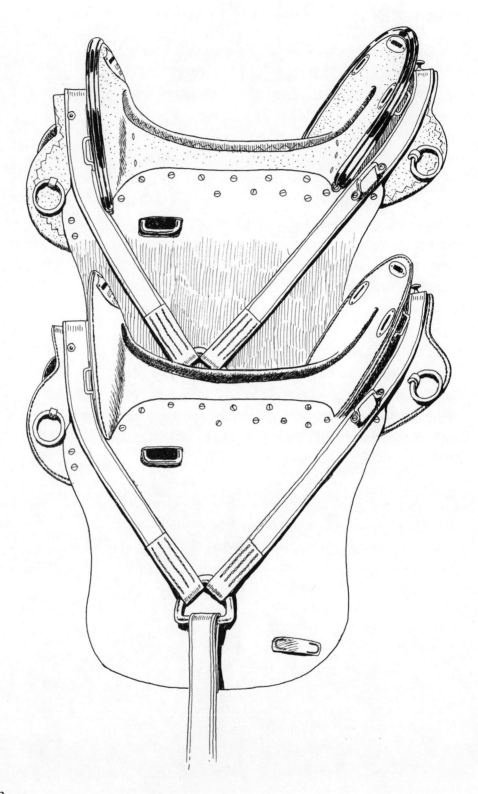

MODIFICATIONS OF THE MCCLELLAN AND WHITMAN

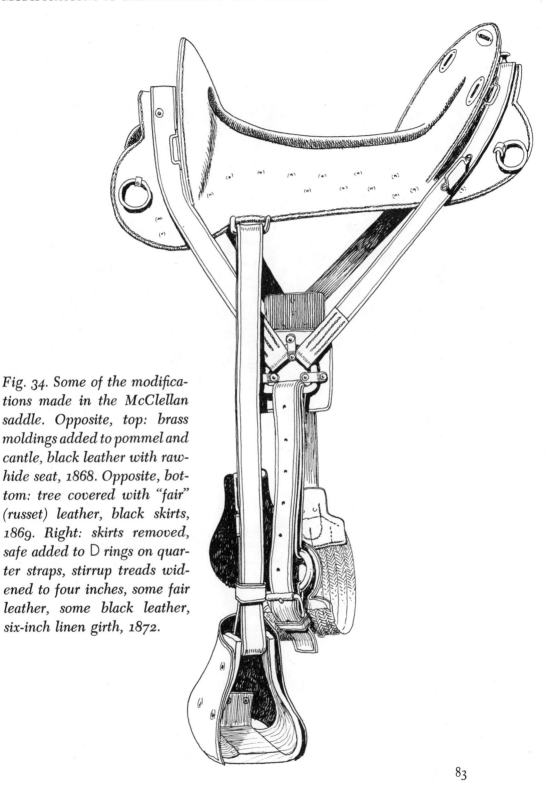

Fig. 34. Some of the modifications made in the McClellan saddle. Opposite, top: brass moldings added to pommel and cantle, black leather with rawhide seat, 1868. Opposite, bottom: tree covered with "fair" (russet) leather, black skirts, 1869. Right: skirts removed, safe added to D rings on quarter straps, stirrup treads widened to four inches, some fair leather, some black leather, six-inch linen girth, 1872.

Fig. 35. Captain, Third Cavalry, in undress uniform, equipped with full pack for service in the field, ca. 1875, near side. In addition to a revolver and a saber he carried the Model 1873 carbine in a pommel loop across his saddle.

A drastic additional change was the removal of skirts, as illustrated in Figure 35. Many saddles that had been covered with leather before the skirts were ordered removed had the screw holes along the sidebars very much in evidence. In my collection is a saddle with these holes visible. Another change, shown with the alterations noted at the right of Figure 34, was the widening of the tread on the wooden stirrups to four inches, almost double the original width. In addition, the old four-and-one-half-inch blue woolen girth was replaced with a six-inch-wide girth with the same buckle ends as before, made of more durable dark-blue linen web. Figures 36 and 37 illustrate these alterations.

In October, 1873, the chief of ordnance recommended to the secretary of war that a board of officers be convened to consider what changes should be made in the "modification of the present horse equipments." The next month the secretary of war directed the adjutant general to convene such a board, made up of cavalry officers who were destined to become prominent in the history of the United States Cavalry in the period of the Indian Wars.

The board sat for five months, examining suggested changes in cavalry equipment, and in May, 1874, submitted recommendations and resolutions to the adjutant general. These resolutions covered changes in most of the cavalry equipments and accouterments, including significant changes in the McClellan saddle. Figure 38 shows the cavalry saddle with the changes and modifications recommended by the board (Ordnance Memorandum No. 18) and adopted by the Ordnance Department later that year. While it is doubtful that any but pattern

Fig. 36. Private, Third Cavalry, in fatigue dress, equipped with full pack for service in the field, ca. 1875, off side.

Fig. 37. Sergeant major, Third Cavalry, in fatigue dress, with full pack for service in the field, ca. 1875, near side.

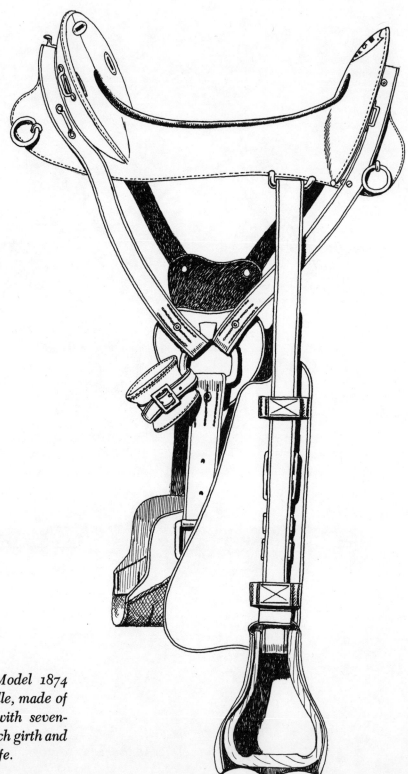

Fig. 38. The Model 1874 McClellan saddle, made of black leather with seven-and-one-half-inch girth and heart-shaped safe.

saddles incorporating these changes were made at the arsenals before about 1876 or later, saddles with these features were designated Model 1874. Figure 39 shows the saddle and horse equipments about 1878.

A few minor changes were made in the basic dimensions of the wood tree, these changes being specifically covered in a series of eleven detailed mechanical drawings, which included six-dimensioned sectional views besides the conventional front, rear, side, top, and bottom elevations. The saddle was to be covered with black leather and without skirts. The seams on the pommel and cantle were to be reinforced with welts of leather, and new heart-shaped safes were to be attached to the quarter straps on both sides and to the girth billet on the off side only. Rings at the ends of the bars were to be of brass, as were the staples used to fasten them to the bars. The quarter straps were to be fastened to a D ring on the off side and to a circular ring on the near side—both rings to be iron and japanned black. The girth was to be widened from the 1872 width of six inches to seven and one-half inches, with a length of twenty-five inches and was to be made of strong blue linen webbing. Stirrup treads were reduced from four to three inches, and sweat leathers were again made an item of issue, after having been abolished shortly after the end of the Civil War. The carbine socket remained the same pattern as that of the original Model 1859, and was to remain this way for another five years before an improvement was made in it. Figures 40 and 41 illustrate these modifications.

Fig. 39. Captain, Third Cavalry, in undress uniform, with the Model 1874 horse equipments and the 1875 officer's rifle for service in the field, ca. 1878, near side.

Fig. 40. First sergeant, Third Cavalry, in campaign dress, with full pack and Model 1874 equipments and accouterments for service in the field, ca. 1878, near side.

Fig. 41. Corporal, Third Cavalry, in campaign dress, with Model 1874 equipments and accouterments and full pack for service in the field, ca. 1878, off side.

MODIFICATIONS OF THE MCCLELLAN AND WHITMAN

Saddlebags were enlarged from the small Civil War type shown in Figure 29 to a larger rectangular shape with pockets ten by twelve inches in size. The surcingle remained the same as on the Model 1859.

In another few years, over General Sherman's protest, another board of officers would recommend the replacement of the McClellan saddle with another pattern, but Sherman, now General of the Army, quashed any major changes in his usual high-handed manner.

The content of General Order No. 76, issued on July 23, 1879, by the adjutant general's office in Washington, was concrete evidence that the service McClellan saddle had lost favor in the mounted service. A board of officers had been convened during the winter of 1878 to consider the equipment of the army in general. The board's remarks and recommendations concerning cavalry saddles are most interesting and reveal the general opinion of the McClellan saddle:

The Saddle

In taking up this subject the Board, while remembering that the McClellan tree has been of great service, is satisfied that a change is now necessary. This conclusion is due in a measure to the experience of the Board, but chiefly to the opinions of a great many officers who are riding saddles of various kinds, many favoring the different forms of the California or Mexican tree, some the English tree, others the Grimsley or the Jenifer. These discriminations against the McClellan saddle cannot be groundless, and are no doubt due to the shape of the tree and the uncomfortable "forked" seat which it occasions.

In seeking to obviate this and other objections the Board has endeavored to find a saddle combining the merits of the various trees now in use.

This, it is believed, has been done in the selection of the Whitman tree, which the Board feels quite confident will meet with general favor in the Army. It is known that a large number of officers, more especially those who have served on the Pacific Coast, are strongly in favor of a horn in the pommel of the saddle. The Board, therefore, recommends that of the first 1,000 saddles manufactured 500 be made with and 500 without horns ... service in the field to determine the final adoption or rejection of the horn as part of the tree.

While retaining the carbine socket the Board has placed a carbine loop in the pommel, believing that to be the most convenient place to carry the carbine on scouts and long marches. In fact, this is the universal custom of horsemen on the frontier, both white and Indian, and in many cavalry regiments company officers permit their men to carry their carbines in a similar way. The Board still deems it best, however, to retain the carbine sling and swivel for use on foot. The Board recommends the carbine socket designed by Sergeant Henry Hartman, First Cavalry, believing that it possesses decided advantages over the present patterns in the facility with which the carbine can be liberated by simply pressing the carbine outward, a great safeguard were a rider to be thrown with his carbine slung.

Figure 42 shows the Whitman saddle with a horn and the frontier-type carbine loop. The "muley" Whitman without a horn and with the carbine loop fastened to the pommel with a buckle strap and stud is shown in Figure 43.

Another change in equipment urged by this board was the adoption of hair cinches instead of linen web girths, to be attached to the saddle by means of a cinch strap instead of buckles. Both of these items are illustrated in Figures 42 and 43. The board also considered it advisable to eliminate the stirrup hoods

Fig. 42. The Whitman cavalry saddle with horn as issued to some cavalry units in 1879 for field trial. The frontier-type carbine loop on the horn was a new feature, furnished in addition to the new Hartman carbine socket shown attached to the quarter-strap ring.

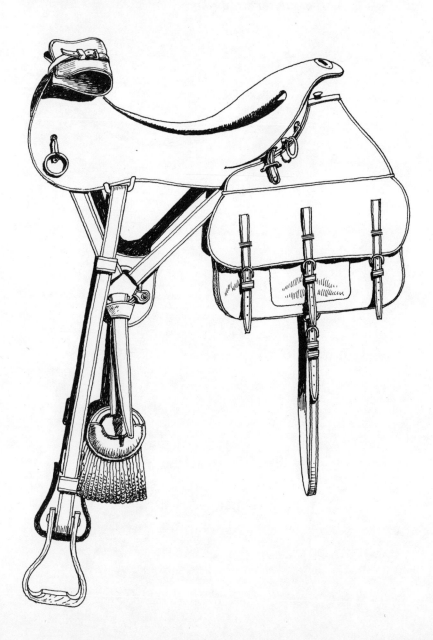

Fig. 43. The experimental Whitman saddle without horn issued in 1879. The saddlebag pattern is the first "California type" issued. The carbine loop is attached to the pommel with a buckled strap and stud.

used since 1859 and to enlarge further the dimensions of the wooden stirrups to allow overshoes to be worn in winter and to give increased strength to the stirrup.

The Hartman carbine socket mentioned above is shown on the Whitman saddle with the horn in Figure 42. Made with a spring-steel core one-sixteenth inch thick and two inches wide and covered with black collar leather, it had a narrow opening at the side through which the barrel of the carbine could slip when pressure was exerted on it by the cavalryman.

The Whitman saddlebags introduced with this saddle were the first of the "California-type" bags. They were made of leather with detachable canvas linings. To steady the bags a strap three-quarters of an inch wide and six feet long buckled to two rings riveted to the bottom of the bags and passed under the belly of the horse. This type of saddlebag is shown in the drawing at the right in Figure 43.

General of the Army Sherman held a tight grip on the purse strings of the army. In his report to the secretary of war concerning the findings of the board that had recommended abandonment of the McClellan and adoption of the Whitman saddle (besides dozens of changes in other equipment), Sherman did a complete turnabout as far as approving changes was concerned from his adamant stand a few years earlier against any changes. He pointed out in this report that the chief of ordnance had on hand for issue forty-two thousand new McClellan saddles (no doubt the recently made Model 1874), enough to supply the wants of the army for ten years. "Nevertheless," he wrote, "it seems to me some of the changes recommended by the Board may be made at

Fig. 44. Private, Cavalry, in campaign dress and equipped with fully packed Whitman equipments issued in limited quantities to cavalry companies for trial in the field, ca. 1881, off side. The "California" horn and carbine loop were new for the military.

once without loss to the Government and yet add to the general efficiency of the Army, whilst others can be 'approved' and the changes can afterward be made gradually as the existing stores on hand are consumed."

Sherman approved the adoption of the Whitman saddle, some with horns and some without, but added a qualifying remark: "Recommend its adoption for experiment, and for general use when the present stock of McClellans is reduced below 20,000." The Whitman saddle with the California horn is shown in Figure 44.

Considerable numbers of Whitmans were made up at the arsenals and issued to some of the cavalry companies for field trial, but the enthusiasm for this change diminished with time, for in 1885 the McClellan was changed again, and some of the changes and improvements planned for the Whitman were incorporated, as illustrated in Figure 45.

Another special board of officers, convened in 1885, made their recommendations for a considerable number of changes in cavalry equipments, including a few minor changes in the McClellan tree, substitution of cincha straps, or latigos, for the former girth billets, a twenty-four-strand hair cinch for the wide linen web girth, and a new type of short boot in which to carry the carbine. This boot had a brass throat to prevent wear by the carbine hammer, and was attached to the saddle with two buckle straps around the quarter strap and the quarter-strap ring and further suspended from the saddlebag stud with a strap riveted to the front part of the boot. These modifications are illustrated in Figures 46 and 47.

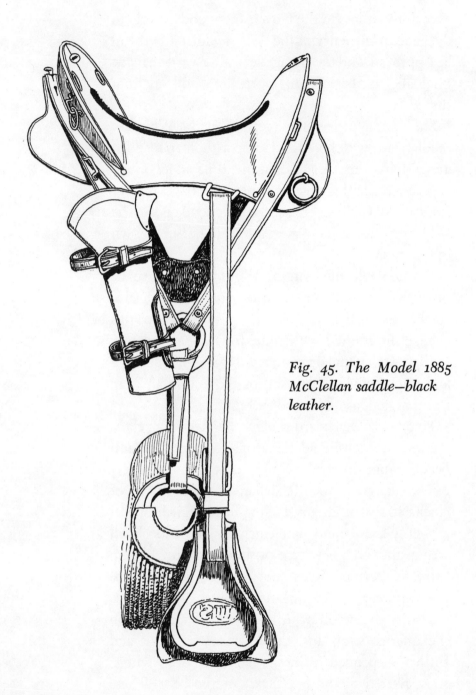

Fig. 45. The Model 1885 McClellan saddle—black leather.

MODIFICATIONS OF THE MCCLELLAN AND WHITMAN

Fig. 46. Sergeant major, Sixth Cavalry, in campaign dress and equipped with full pack for field service, ca. 1885, near side.

Fig. 47. Private, Sixth Cavalry, in campaign dress and with full equipment for service in the field, ca. 1885, off side.

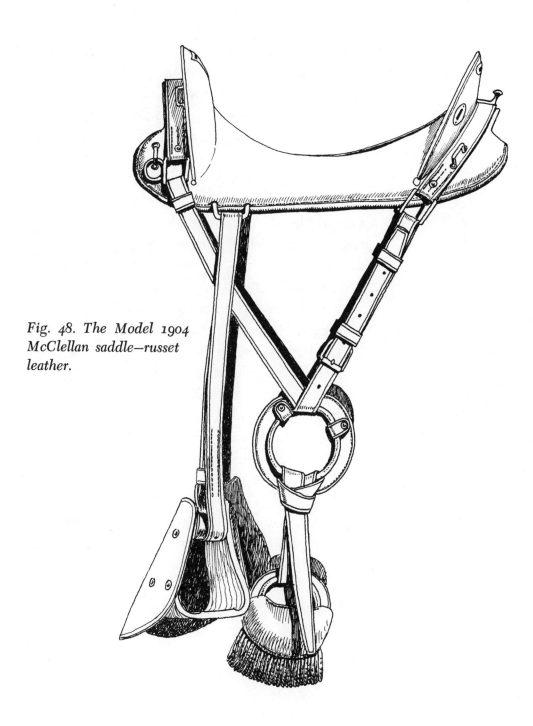

Fig. 48. The Model 1904 McClellan saddle—russet leather.

The Model 1885 McClellan shown in Figure 45 was to be the last United States saddle covered with black leather. The 1902 Regulations for the Uniform of the Army changed all leather parts of the uniform from black to russet, with the exception of dress shoes and boots. That was also the case with all leather equipment and accouterments prescribed for both officers and enlisted men, and naturally saddles followed suit.

Figure 48 shows the Model 1904 McClellan, which was the last production model of this long-lasting army saddle. There was another modification more than two decades later, but the tree, girth, and stirrup leathers remained as they were settled on in 1904.

Work had started on the changes evident in this first russet saddle a number of years earlier, for there exist black McClellans made to the same specifications and dimensions as the russet Model 1904 except for the color of the leather. I have such a black saddle in my collection, as well as the standard russet specimen.

Characteristics that distinguished the Model 1904 from previous models, in addition to its russet color, are the two-piece quarter straps, the rear ones adjustable by means of large bronze roller buckles. The bottom edges of the sidebars are almost perfectly straight instead of curved, as were all previous models, and all metal fittings have a dull bronze finish. The quarter-strap safes have a new circular shape with larger rings, and the stirrups are larger and heavier.

MODIFICATIONS OF THE MCCLELLAN AND WHITMAN

Like the Model 1885 McClellan, the Model 1904 McClellan was made in both cavalry and artillery models, the only difference being the addition of heavy leather straps and buckles at both pommel and cantle, to which parts of the artillery harness were secured.

8 CAVALRY BOARD SADDLES, 1912-16

A cavalry equipment board of officers was convened in 1910 to make recommendations for a high-quality cavalry saddle and related equipment. Again there had been an almost overwhelming feeling among cavalry officers that the McClellan was lacking in many ways as the best saddle for the mounted service. With the replacement of the McClellan uppermost in mind, members of the board set about their task in a serious and thorough manner.

The board had before it more than three thousand documents representing the opinions of hundreds of officers in the mounted service. It also had for inspection horse equipments from all the major European powers, Japan, and Mexico. To supplement the equipments were voluminous comments by American attachés in London, Berlin, Paris, St. Petersburg, Vienna, Tokyo, Peking, Buenos Aires, Santiago (Chile), and Lima (Peru), including photographs, drawings, and compilations of statistics. The board augmented its own knowledge with consultations involving key personnel of experienced and and successful business firms and manufacturers engaged in the production of first-class saddlery and related equipment. In its actual work of design and development of equipments, the board had at its disposal the resources of personnel and plant of the Rock Island (Illinois) Arsenal and the active support and assistance of the chief of ordnance.

In September, 1912, the first photographs and descriptions of the 1912 cavalry equipment were published in the *U.S. Cavalry Journal* in a comprehensive article by Captain Edward Davis of the Thirteenth Cavalry, a member of the board. Figure 49 illustrates the 1911 saddle with authorized equipments. Figure 50 shows the officer's and service saddles and a view of the wood-and-metal tree on which both saddles were constructed.

Fig. 49. Private and sergeant of cavalry, ca. 1912, off side and near side. They are wearing the uniform authorized in 1911, and their Model 1904 McClellan saddles are fully packed with regulation equipment.

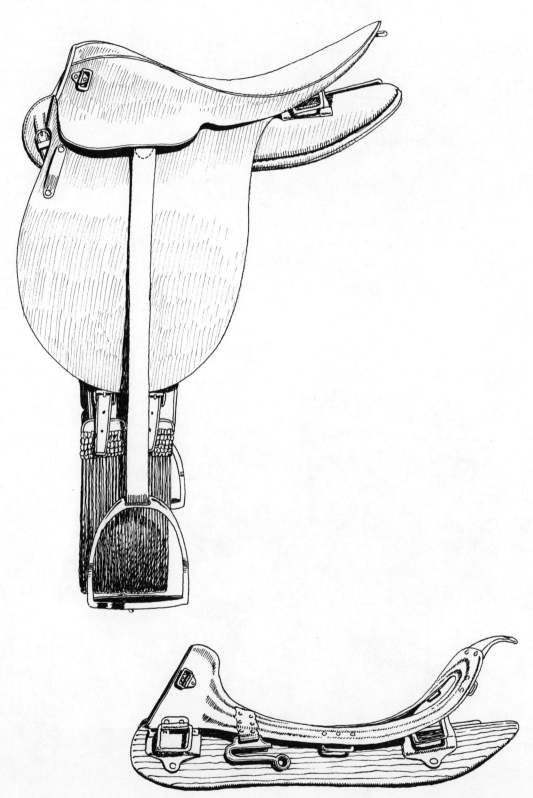

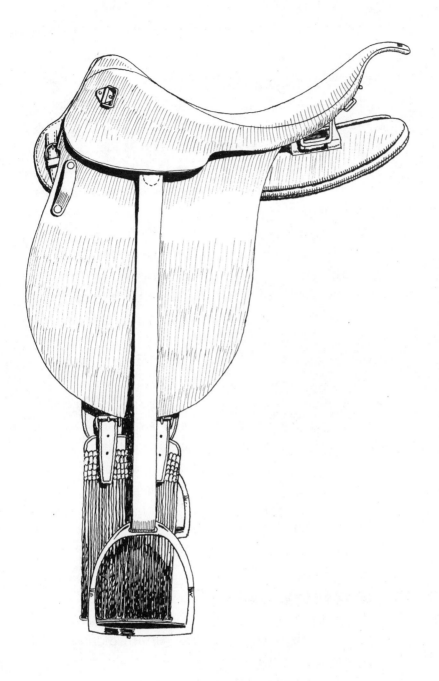

Fig. 50. The Model 1912 experimental cavalry saddle. Opposite, top: officer's saddle. Opposite, bottom: the Model 1912 tree with steel seat and self-adjusting hinged sidebars. Above: the Model 1912 trooper's service saddle.

The outstanding feature of this saddle was, of course, its self-adjusting bars. The stamped sheet-steel part of the tree was joined to the wooden bars by means of four rugged hinges, and made the saddle adjustable to any horse's back. The Model 1912 saddle, as it was officially designated, was not by any means the first self-adjusting saddle to be issued for trial in the United States Cavalry. The Campbell saddle of 1855 had had the same feature, as shown in Figure 22. The next year, 1856, consideration had again been given to the saddle invented by Lieutenant W. E. Jones, which employed a hinged sheet-metal tree to which the sidebars were fastened. But technical inadequacies of the period made the Jones saddle impractical to manufacture with sufficient strength to hold up under severe cavalry use.

In appearance the Model 1912 saddle resembled several of the European cavalry saddles. The shape of the pommel was much like that of the German cavalry saddle, and the cantle was similar to that of the French officer's military saddle, while the dip, or curve, of the seat was much like that of the British service saddle.

Under the seats of each of these saddles was a ground seat of heavy sole leather, and under this were two strips of the best English straining web, all supported by the steel frame shown at the bottom left of Figure 50. The long sidebars afforded ample support for the cantle pack, and the bars were turned up sufficiently on the ends to prevent them from gouging into the horse's back. The first versions of the Model 1912 service saddle had a cantle peak, or pack support, that folded back under the cantle when not in use. But on the production run of saddles issued by

CAVALRY BOARD SADDLES

the Rock Island Arsenal for extensive tests over a number of years in the field, this cantle support was made rigid and was covered with the seat and jockey leather, as shown in Figure 50.

The stirrup was of nickel steel with a dark, sanded, oxidized finish, and required only wiping with an oily cloth to keep it in top condition; polishing was prohibited. This was the same stirrup used on the

Fig. 51. First sergeant and private with Model 1912 equipments, off side and near side. Both are wearing the 1911 regulation uniform with standing collar and the Model 1912 leather leggings. The sergeant is carrying the noncommissioned officer's record case over his bandoleer, according to regulations.

Fig. 52. Captain, Fifth Cavalry, in summer service dress, with Model 1912 equipments and full pack for duty in the field, ca. 1914, off side.

CAVALRY BOARD SADDLES

McClellans made for the field artillery. The equipment board contended that the steel stirrup permitted better horsemanship and more comfortable riding, lasted longer, had less bulk, weighed less, cost less, and looked better than the hooded wooden issue stirrup.

Attached to the underside of each sidebar by means of leather pockets was a felt pad one-half inch thick. The purpose of the pads was to create friction to help keep the saddle in place without tight cinching and to afford a cushioning effect to help relieve the horse from the jolts and jars of the weight of the rider and the pack. The removable pads also allowed the building up or cutting away of the bulk of the bars in the treatment of cuts, wounds, or sores while on the march.

The metal bars on which the stirrup leathers were hung were of the safety-loop pattern.

The girth was a thirty-five-strand braided-cotton olive-drab sash cord with buckles at each end. A swivel buckle in the center of the girth on the underside allowed the attachment of a strap to be used for steadying the rifle and saber when they were carried.

A unique system of girth adjusters, through which the girth billets passed, allowed the girth to be adjusted to horses with different shapes of barrel and also allowed the position of the saddle to be moved forward or back on the horse's back as necessary.

The Model 1912 saddle was two pounds fourteen ounces heavier than the McClellan. But the use of a hair pad lighter in weight than the folded blanket used with the McClellan made this difference nil.

In 1912 the McClellan saddle, complete with stirrups, cinch, and coat straps, was listed in the March

1, 1910, Ordnance Price List at $22.40; the Model 1912 saddle was estimated to cost $21.00.

The Ordnance Board approved a complete new set of equipments to accompany the new saddle: a combination halter and bridle with bit and bridoon; a rifle carrier, or bucket, and an entirely different

Fig. 53. Captain, Fifth Cavalry, in service uniform, with Model 1912 equipments required for garrison duty under arms, ca. 1914, off side.

cartridge belt with a padded ring through which the muzzle of the rifle was thrust when it was carried afoot or mounted; a new saber with a straight blade designed primarily for thrusting (the Model 1913 saber designed by Lieutenant George S. Patton, Jr., destined to gain fame as "Old Blood and Guts" during World War II); a new saber carrier—one for officers and one for the service saber for both officers and enlisted men; a combination picket pin and entrenching shovel, ax, and pick; an entrenching tool carrier; a new type of feed bag; pommel pockets; and ration bags that converted from cantle bags to a shoulder pack for the dismounted cavalryman. The 1912 equipments are shown in Figures 51 to 53.

Hundreds of sets of the Model 1912 equipments were made up at the arsenals and issued to cavalrymen in the field. Issues of the new equipment were made as late as 1918. Then, for some reason, experiments were begun with new designs, and the 1912 saddle and most of its related accessories went into storage. Very little of this equipment exists today. One officer's saddle and one service saddle are on display at the Quartermaster Museum at Fort Lee, Virginia.

One of the new designs resulted in more extensive field trials by various troops of the several cavalry regiments. Apparently the equipment board was not satisfied with the self-adjusting Model 1912 saddle, for another, without hinged bars and with a partly open seat, was illustrated and described in the April, 1917, issue of the *U.S. Cavalry Journal*. A surviving specimen is on display at the Quartermaster Museum in Fort Lee, Virginia. The drawings in Figure 54 were made from that saddle. Designated the Model

UNITED STATES MILITARY SADDLES

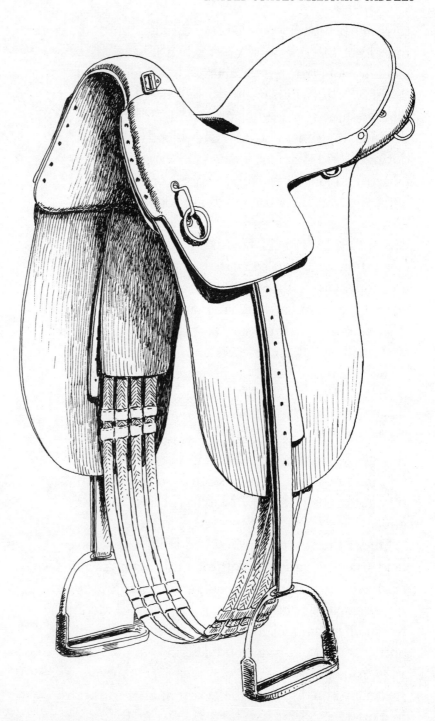

CAVALRY BOARD SADDLES

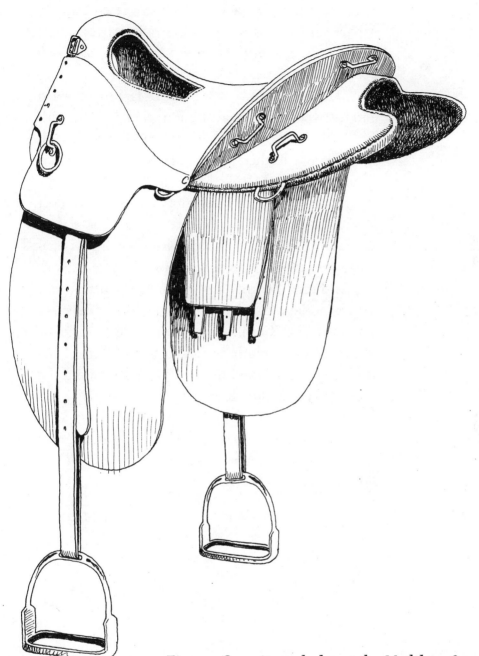

Fig. 54. Opposite and above: the Model 1916 experimental cavalry saddle, issued in limited quantities to some cavalry regiments for field test on the Mexican border.

1916 experimental saddle, it closely resembles the Model 1912, except for the solid leather-covered bars and the opening at the forward end of the seat—an opening a little larger than that common to most stock saddles. The tree was made of laminated wood and basswood reinforced at pommel and cantle with steel arches. Like the McClellan, it was covered with rawhide and leather. Arsenal tests showed it to be stronger than the McClellan.

After a number of satisfactory field tests at Fort Riley, Kansas, in the early part of 1917, more than one hundred sets were made at Rock Island Arsenal and issued to some cavalry troops for more exhaustive field tests. Seventy sets of the new experimental service saddle with related equipments were issued to Troop D, Fifth Cavalry. The commanding officer of this troop was enthusiastic about the comfort of the saddle. His report said in part: ". . . after riding the [new] service saddle fourteen consecutive days, . . . I have arrived at the following conclusions: It is more comfortable than the McClellan; it is easier on the horse on account of the high pommel arch and the position of the bars."

9 OFFICERS' SADDLES, 1917-26

At the same time the cavalry equipment board was developing and testing the service saddle shown in Figure 54, it came up with another saddle designed especially for officers, designated the Model 1916 officer's saddle. This saddle later received official permanent designation as the Model 1917 officer's saddle. It is shown in Figures 55 and 56 with the pommel and cantle pockets, Model 1917.

This saddle was very much like the officer's saddle shown in Figure 50, except that the cantle was slightly lower at the sides, the bars were solid, and it was fitted with the Model 1916 steel stirrups. In addition, the pommel bags fit into a special metal socket mortised into the pommel. Bags used on the enlisted man's 1916 saddle were attached in a different manner. The Model 1917 officer's saddle was the official field saddle for cavalry officers until the introduction

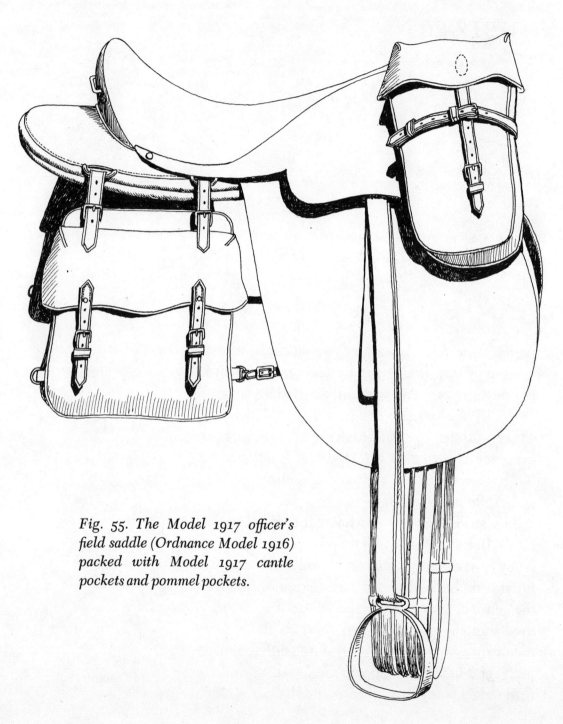

Fig. 55. The Model 1917 officer's field saddle (Ordnance Model 1916) packed with Model 1917 cantle pockets and pommel pockets.

OFFICERS' SADDLES

Fig. 56. Captain, Second Cavalry, in service uniform, with the Model 1917 officer's field saddle packed for service in the field, ca. 1917, near side. His bridle is the Model 1912 combination halter-bridle. He is uniformed and equipped for duty in France.

of the Phillips cross-country saddle in the 1930's. Figure 57 shows a cavalry officer riding the Model 1917 saddle.

Figure 58 shows the Model 1916 officer's training saddle developed by the same cavalry equipment board responsible for the design of the Model 1912, the Model 1916 experimental saddle, and the Model 1917 officer's saddle. Used at the Cavalry School at

Fig. 57. Company officer, Cavalry, in field dress, riding the Model 1917 officer's field saddle with full pack for service in the field, ca. 1931, off side. He is armed with the service saber and the Model 1911 Colt automatic pistol. He is running at the heads in a field training exercise.

OFFICERS' SADDLES

Fort Riley strictly for training purposes, it continued to be available for purchase by officers from the Ordnance Department until the Phillips cross-country saddle made it obsolete about 1936. It is shown as a standard item of saddlery in the 1930 edition of *The Quartermaster Handbook*. Figures 59 and 60 illustrate the saddles used for overseas and continental United States service in 1917.

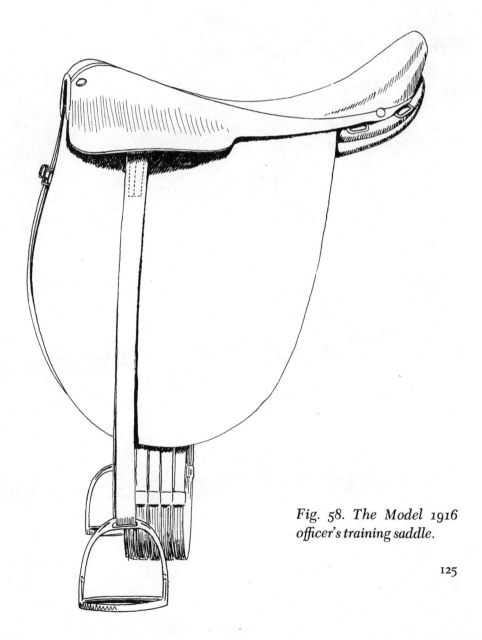

Fig. 58. The Model 1916 officer's training saddle.

Fig. 59. Corporal, Second Cavalry, in field dress, with steel helmet for service in France, ca. 1917, near side. His Model 1904 saddle is packed for service in the field, and he carries the regulation gas mask over his right shoulder.

OFFICERS' SADDLES

Fig. 60. First sergeant, Third Cavalry, in summer field dress, on the Model 1904 McClellan saddle, with full pack for service within the continental United States, ca. 1917, off side. His bridle is the Model 1909 cavalry bridle. He is armed with the 1911 .45-caliber automatic pistol, the Model 1903 rifle, and the Model 1913 cavalry saber.

UNITED STATES MILITARY SADDLES

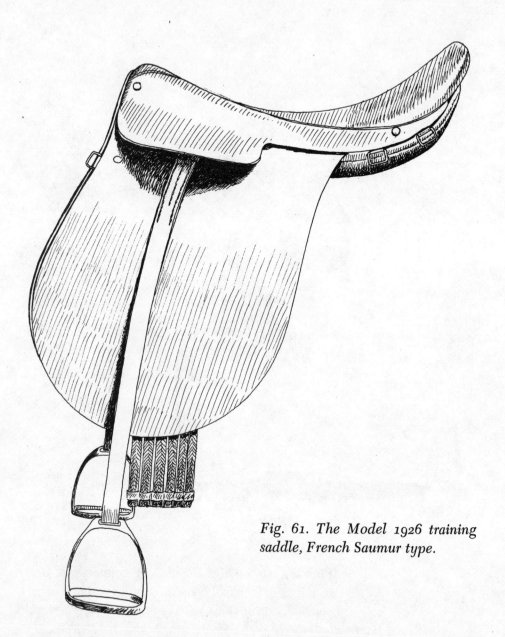

Fig. 61. The Model 1926 training saddle, French Saumur type.

OFFICERS' SADDLES

For quite a few years during the early part of the twentieth century selected American cavalry officers were sent to the Cavalry Training School at Saumur, France. There they rode one particular style of saddle, originally made by Marquis Saddlery for the French cavalry officers. For a number of years these saddles were privately purchased by American cavalry officers from Marquis and later from Hermès Saddlery, which took over the Marquis Saddlery about 1929. This saddle, shown in Figure 61, was copied by the United States Ordnance Department, was designated the Model 1926 training saddle, Saumur type, and was made available to army officers for purchase.

10 THE LAST McCLELLAN AND THE PHILLIPS SADDLE, 1928-43

Figure 62 shows the last modification of the Model 1904 McClellan saddle, designated the Model 1928. This modification was standard through the beginning of World War II, when the last of the horse cavalry was dismounted and the word cavalry became a symbol for the reek of gasoline and oil.

In this modification the quarter straps were removed, double girth billets were sewed to the leather tree covering, and collar-leather skirts were added. It now closely resembled the old army saddle that had started off with skirts almost identical to these sixty-nine years earlier. New olive-drab web girths replaced the hair cinches that had become standard with the McClellan in 1885, and new-style stirrup leathers were installed.

The Model 1928 was also made in artillery style for individual mounts, and differed from the cavalry model only in the length of cantle coat straps. Thousands of surplus Model 1928 modified McClellans are still in use all over the world and are favorite saddles, along with many older, unaltered Model

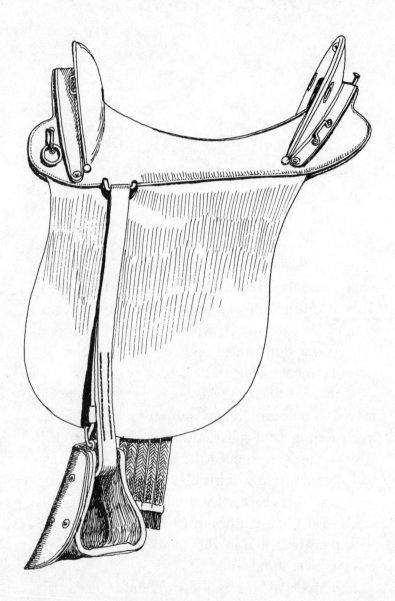

Fig. 62. The Model 1928 modified McClellan, the last model issue cavalry saddle of the United States Army.

THE LAST MCCLELLAN AND PHILLIPS

1904's, among endurance riders in the United States and Australia. The Model 1928 is shown with full pack for service in the field in Figures 63 to 65.

Fig. 63. Private, Seventh Cavalry, in summer field dress, ca. 1931, off side. He rides the Model 1928 modified McClellan saddle with full pack for service in the field. His saber is drawn for a run at the heads in the practice course.

Fig. 64. Private, Seventh Cavalry, in summer field dress, ca. 1931, near side. He rides the Model 1928 modified McClellan saddle with full pack for service in the field. He is armed with the Model 1903 rifle, the .45-caliber Model 1911 Colt automatic pistol, and the Model 1913 cavalry saber.

THE LAST MCCLELLAN AND PHILLIPS

From the time of the punitive expedition into Mexico until his death not too many years ago, an American cavalry officer, Colonel Albert E. Phillips, was closely associated with all major developments in military pack and riding saddles. It was Colonel Phillips who developed the famed Phillips packsaddle. The Phillips replaced the old Spanish-type aparejo used from the Indian-fighting days of the old army until the modified aparejo-type packsaddles provided the means for transporting all supplies not hauled in baggage wagons or trucks during and after World War I. It was also Phillips who designed the officer's saddle that was officially approved by the army in 1936 and designated the Model 1936 officer's cross-country saddle. It replaced the Model 1917 as a field saddle for officers, and was highly acclaimed by cavalry officers and civilians alike as a great all-around saddle for general use. Figure 66 shows this saddle, stripped and with the Model 1936 pommel and cantle pockets—slight modifications of the Model 1917 saddle accessories.

Fig. 65. Private, Fourteenth Cavalry, in combat dress, ca. 1941, off side. He rides the Model 1928 modified McClellan with full pack for service in the field. He is armed with the Garand M-1 semiautomatic rifle and the Colt automatic pistol.

THE LAST MCCLELLAN AND PHILLIPS

The extended sidebars of the Model 1917 officer's saddle disappeared. Colonel Phillips pointed out that, although it had been the general belief that the greater bearing surface of the long sidebars distributed the weight more correctly over a greater portion of the horse's back, this was incorrect, for the anatomical structure of a horse made it inadvisable—in fact, detrimental—for any weight or interference to be placed behind the horse's center of motion, which is the fifteenth or sixteenth dorsal vertebra. Phillips proved that an overextended surface of the sidebars was a positive deterrent and was the cause of many disabled horses in the field. The sidebars of the Phillips saddle were no longer than the rear of the cantle, and a notable feature was the detachable cantle-roll support, shown in Figure 66 with the pommel and cantle pockets in place.

The pommel pockets attached to the saddle by means of a stud fitting into a special metal socket mortised into the tree, identical in principle to the method of attaching pommel pockets to the Model 1917 field saddle. This socket is shown on the stripped saddle in Figure 66. The cantle-roll-support feature makes it possible to use this saddle for both training and field use, as Phillips pointed out in an article in the May–June, 1935, issue of the *U.S. Cavalry Journal*. This saddle had a nineteen-and-one-half-inch bearing surface and an eighteen-and-one-half-inch seat, and was made in one size only, thus effecting a great savings in manufacturing cost. The old Model 1917 field saddle had a twenty-five-inch bearing surface and a nineteen-inch seat. Skirts covered inner skirts with padded rolls for the rider's security, but there was enough flexibility even with the roll that

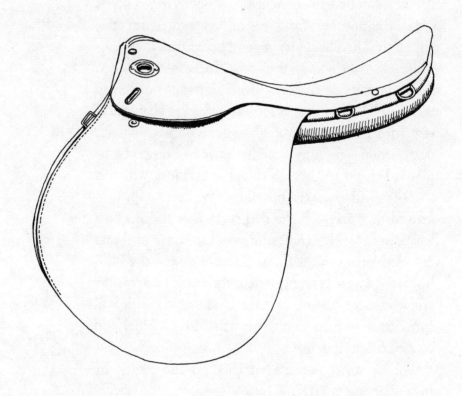

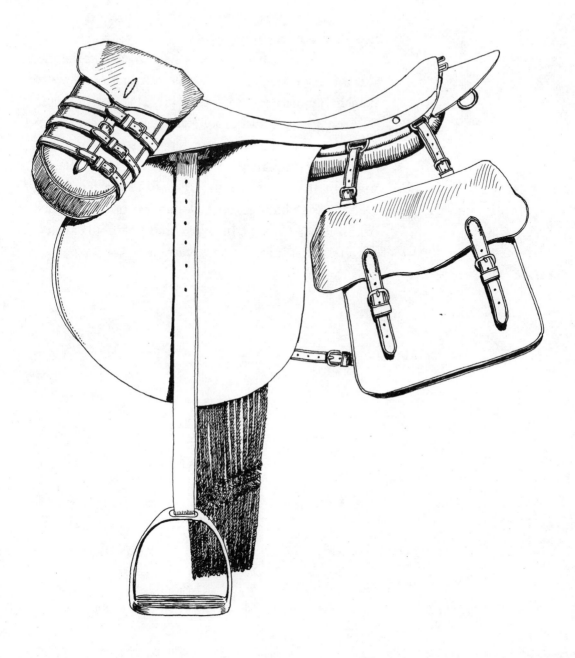

Fig. 66. The Model 1936 Phillips officer's cross-country saddle. Opposite: the bare saddle. Above: the saddle with Model 1936 pommel and cantle bags.

there was no interference with the motion of the horse's shoulders at any gait.

Two types of two-buckle girths were available. One was a Fitzwilliam type, made of woven mohair with detachable leather reinforcing pieces at the buckles. The other was made of mohair cord with two buckles but no leather. Hard-woven felt pads were used instead of the mohair pad. The price of this saddle, as made by the Jeffersonville (Indiana) Quartermaster Depot, was about seventy dollars without pommel or cantle bags. It was the official saddle for all cavalry officers up to the end of horse cavalry.

11 MISCELLANEOUS SADDLES

In addition to the cavalry saddles illustrated and described so far in this book, a number of other saddles should be described to cover all those issued by the United States Army to its various arms.

Figure 67 shows at the left the artillery driver's saddle that was standard equipment for field artillery in 1832, and possibly as early as 1808, when field artillery came into its own. Patterned after the standard military saddle in use in this country in the early nineteenth century, this saddle was built on an English-type tree and covered with heavy collar leather. The same saddle is shown in the comprehensive scale drawings that accompany Alfred Mordecai's *Treatise on Artillery*, published in the early 1840's.

The billet shown on the pommel square was for hitching the saddle to the hame straps on the collar to prevent the saddle from slipping back on the broad-backed, often low-withered artillery draft horses. A crupper attached to the staple set just above the cantle arch prevented the saddle from sliding forward. The girth was made of folded leather, and the stirrups were cast brass. This type of saddle was in constant use and issue until the Grimsley artillery saddles were issued just before the Civil War. It is shown on the rear-wheel horse in Figure 68.

The saddle at the right in Figure 67 is the Model 1832 artillery valise saddle with valise, but with the

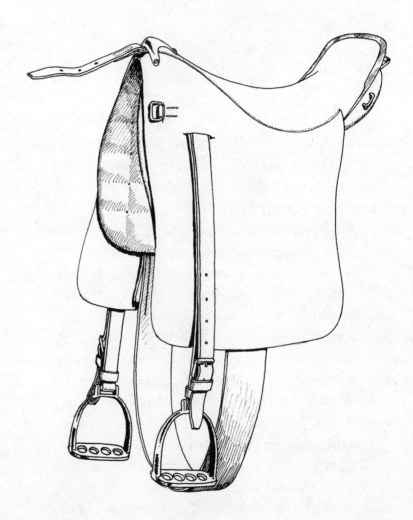

MISCELLANEOUS SADDLES

valise straps slack instead of being buckled tight as they would have been in use. This small saddle was never intended to be used as a riding saddle but was used only on the off-side horse of an artillery team to carry the driver's belongings, packed in the heavy leather valise, which was eighteen inches wide and a little more than seven and one-half inches in diameter.

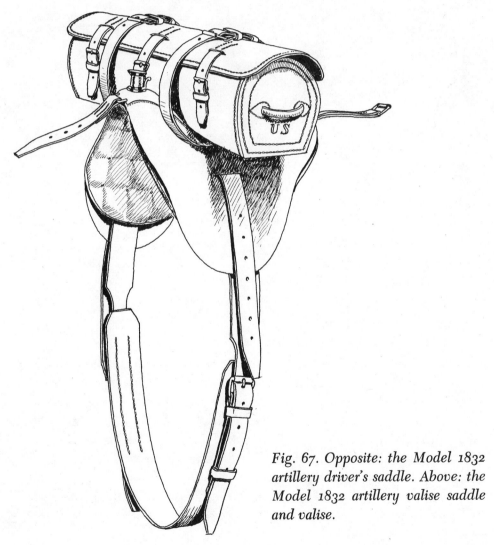

Fig. 67. Opposite: the Model 1832 artillery driver's saddle. Above: the Model 1832 artillery valise saddle and valise.

Fig. 68. Corporal of Light Artillery, wheel-horse driver with the near-wheeler of a field-gun team, ca. 1840. The harness and the driver's saddle on the wheel horse are the pattern of 1832. The harness for the off-wheeler was identical except for the saddle, which was the valise saddle shown in Fig. 67. Each off horse in the six-horse team that pulled the gun and caisson carried a valise saddle, and each near horse was fitted with a driver's saddle. The same type of harness was used with the Grimsley artillery saddles shown in Fig. 69 throughout the Civil War and until late in the 1890's, when McClellans with artillery fittings were adopted for all artillery teams.

MISCELLANEOUS SADDLES

Both driver's and valise saddles were made with padded underskirts to protect the horses' sides from the girth buckles. The valise saddle was made without stirrups or stirrup straps, but had a heavy leather billet that was used to attach to a trace carrier that held the artillery traces in place and prevented them from sagging low enough to allow the horse to step over them.

Figure 69 shows the Grimsley artillery saddles that were standard equipment for field artillery (and for Gatling gun teams and Hotchkiss rapid-fire guns during the Indian Wars) from 1859 or a little earlier until the late 1880's or early 1890's, when field-artillery teams were fitted with McClellan saddles for both drivers and valises.

Very similar to the Grimsley dragoon saddle, the Grimsley artillery driver's saddle was built on a wooden tree covered with rawhide, then black leather and with a quilted seat stuffed with curled hair. Only one set of skirts was used on both driver's and valise saddles, the inner skirt having been dispensed with for some reason. The driver's saddle, shown at the left of Figure 69, was fitted with a billet at the pommel to secure the saddle to the hame straps on the collar and with a large staple at the rear of the cantle to which the crupper was buckled. Stirrups were cast brass, smaller than those on the dragoon saddles, and with a roughened tread instead of a pierced tread.

Both driver's and valise saddles had pommels and cantles bound with brass molding. The girth on the driver's saddle was made of two folded thicknesses of light bridle leather stitched down the middle as shown. The valise saddle girth was made of heavy leather and was attached to the saddle skirt, as was the girth billet.

The trace-loop billets on the valise saddle were fastened directly to the tree and passed through slits in the skirts to hang on the outside of the skirts. On the driver's saddle these trace-loop billets were fas-

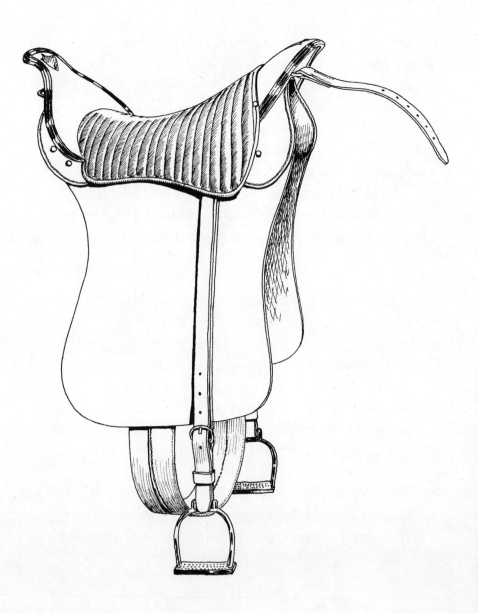

MISCELLANEOUS SADDLES

tened to the tree in a similar manner but hung under the skirts, where they would not chafe the driver's legs.

Exactly the same pattern of valise was used with this model valise saddle as was used with the earlier one shown in Figure 67. The valise straps used to secure the valise to the saddle are shown in front of

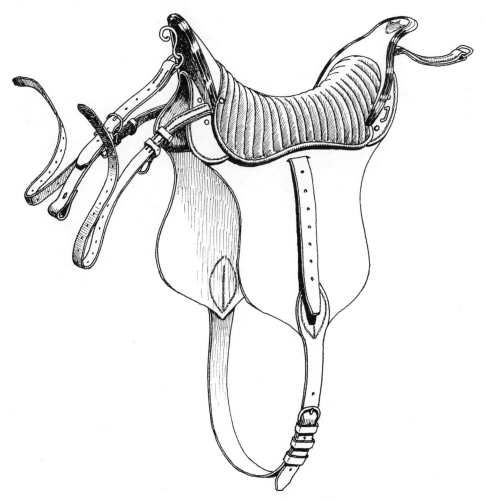

Fig. 69. Opposite: the Model 1859 Grimsley artillery driver's saddle. Above: the artillery valise saddle. The same type of harness, with minor variations as shown in Fig. 68, was used with the Model 1859 driver's and valise saddles.

the face of the pommel, unfastened. When the valise was hung over the seat of the valise saddle, the straps passed over the top of the valise, and their billets passed through the brass staples on the bars of the saddle behind the cantle and buckled to the front part of the valise straps. Two stationary loops sewed to the top cover of the valise kept the valise straps from sliding out of position. These two saddles were drawn from specimens in my collection, and the specifications are quoted from the 1863 edition of the army's *Ordnance Manual*.

Another army issue saddle of the same period is shown in Figure 70. This is the driver's saddle used on army four- and six-horse and four- and six-mule jerk-line teams that pulled the thousands of supply and baggage wagons used just before, during, and after the Civil War. The jerk-line driver used this saddle on the near-wheel animal with the side straps of the breeching running over the skirts of the saddle and under the stirrup leathers to the loop on the hames where they were fastened. The saddle is shown on the near-wheel mule in Figure 71.

This saddle was built on a wooden Morgan tree covered with rawhide and with the barest of rigging, as shown in Figure 70. The girth, sixteen inches long and four inches wide, was made of woven hair. The fenders were removable, being held to the stirrup leathers with two loops riveted to the backs.

This saddle was in use in all branches of the army for their wagon trains until the 1880's or 1890's, when the baggage wagon and its jerk-line teams were replaced by the escort wagon drawn by teams driven with lines.

MISCELLANEOUS SADDLES

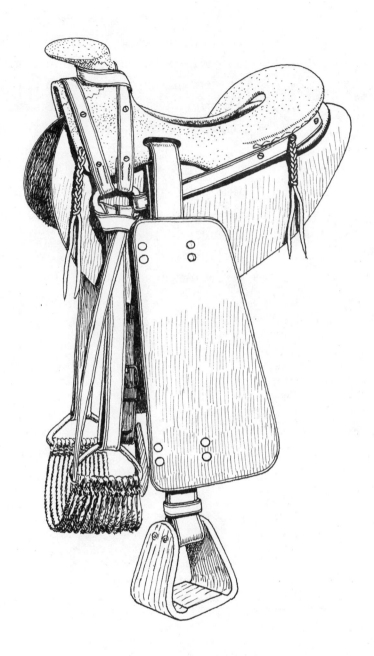

Fig. 70. A driver's saddle for four- and six-horse and four- and six-mule jerk-line teams for army baggage and supply wagons. This saddle was used from 1859 through the 1880's, when jerk-line teams were replaced with teams driven with lines.

UNITED STATES MILITARY SADDLES

Before the adoption of the driver's saddle built on the Morgan tree shown in Figure 70, the army used a driver's saddle almost identical to the Model 1832 artillery driver's saddle shown in Figures 67 and 68. Descriptions of this saddle in quartermaster specifications for harness used on four- and six-mule and horse teams indicate little or no difference from the 1832 artillery model. Civilian wagoners used practically the same saddle on freight teams throughout the United States. In the East during the eighteenth and nineteenth centuries the heavy teams that pulled the huge Conestoga wagons used a similar type, differing mainly in that it had longer skirts. Specimens

MISCELLANEOUS SADDLES

of this saddle can be found in several museums in Pennsylvania and other eastern states.

About 1913 the Quartermaster Department made up a quantity of modified McClellan saddles with two cinches and a metal horn attached to the pommel. This modification was designated the Model 1913 mule riding saddle and was issued to packers who rode saddle mules with the army pack trains. Figure 72 shows this saddle as it was issued, made on a modified Model 1904 tree. The quarter straps were arranged as shown, with two rigging rings and a connecting strap to accommodate two girths, which

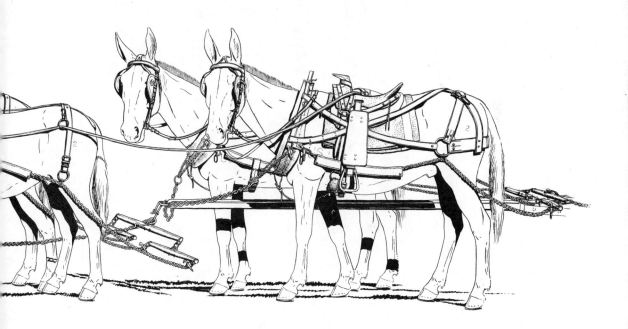

Fig. 71. *A six-mule jerk-line team hitched to an army supply wagon. The driver's saddle on the near-wheeler (above) was in constant use with army freight and supply wagon teams for fifty years, from the late 1850's through the turn of the century.*

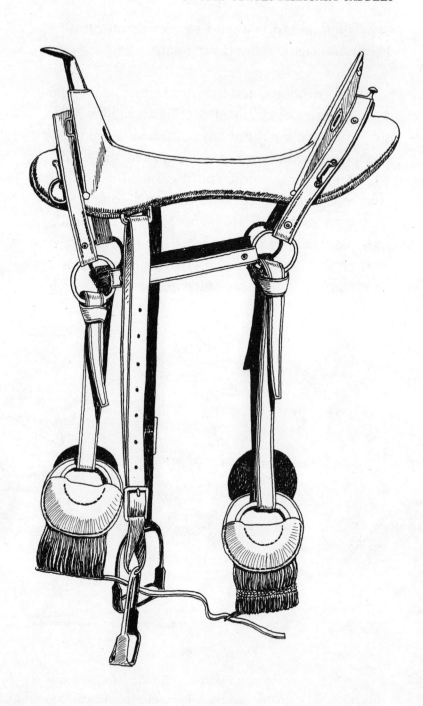

Fig. 72. The Model 1913 mule riding saddle.

MISCELLANEOUS SADDLES

were buckled together to keep them from slipping too far out of place. Breeching and breast harness was used on most mules, being attached to the saddle by means of the standard rings and staples on both pommel and cantle ends of the bars. Many of these old saddles still exist and are being used, but few people seem to be familiar with their original use.

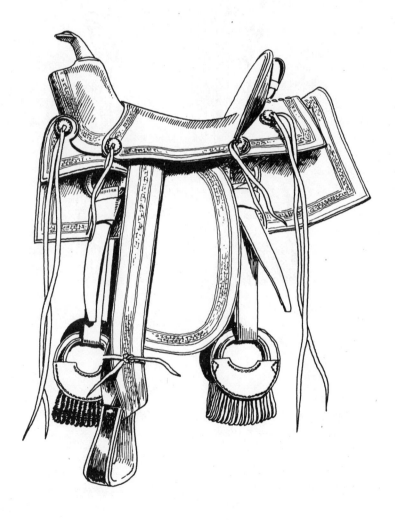

Fig. 73. A full-rigged packer's riding saddle, ca. 1917.

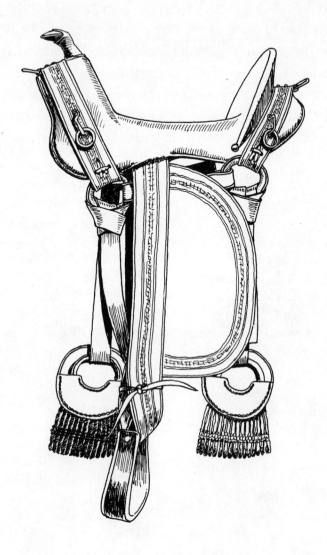

Fig. 74. A skeleton-rigged packer's riding saddle, generally used on mules, ca. 1917.

MISCELLANEOUS SADDLES

About 1917 the Quartermaster Department issued the saddles shown in Figures 73 and 74 to cavalrymen attached to remount depots whose duty it was to break the thousands of horses being purchased for the army. There had been so many complaints about the inadequacies of the McClellan saddle when it came to staying on a pitching green remount horse that the Quartermaster Department adopted these two stock saddles for the horse breakers. Later they were issued to packers for use on the riding animals with the pack trains. They were an item of issue as late as 1944; they are part of the equipment carried in the *Quartermaster Supply Catalog* of that year. No doubt the skeleton-rigged packer's saddle was meant for use on saddle mules, for metal loops for breast harness and breeching are provided at both pommel and cantle.

INDEX

Artillery, United States Army: 38, 115; McClellan saddle models for, 104–105, 113, 145; driver's saddle, 1832, 141–48, 150; valise saddles, 142–48; Grimsley saddles for, 145; driver's saddle for jerk-line team, 148, 150–51

Brady, Mathew B.: 78

Campbell, Daniel: 50; *see also* Campbell cavalry saddle of 1855
Campbell cavalry saddle of 1855: 50–51, 54, 57; issued to First and Second Cavalry Regiments, 50, 54, 64–65; replaced by Grimsley saddle, 1856, 54
Cavalry Training School, Saumur, France: 129
Civil War: 41, 49, 79, 89, 93; Grimsley saddle used in, 46; Hope saddle in, 54, 57; McClellan saddle in, 70, 73, 75, 78; Jenifer saddle in, 75, 78; and jerk-line teams, 148
Confederate Museum, Richmond, Va.: 57, 78
Cooke, Major Phillip St. George: 41
Coxe, Tench: 3–4

Davis, Captain Edward: 108
Davis, Jefferson: 49
Dragoon saddle, 1841 model: 6, 24, 28, 31, 37
Dragoon saddle, 1833 model: 17–18, 22, 24, 37, 51; replacement of, 22, 24, 28
Dragoon saddles, 1792–98: 3
Driver's saddle, jerk-line team: 148, 150–51

Eighth Cavalry, 1866: 79
English-type saddle: *see* dragoon saddle, 1833 model

Fifth Cavalry: 120
First Dragoons, 1846: 38, 41
First Regiment of Cavalry, 1855: 49–50, 54; Campbell saddles issued to, 50, 54, 112
Fort Riley, Kans.: 120, 124–25
Fort Riley Museum, Kansas: 19
Fort Sill Museum, Oklahoma: 57

INDEX

Grant, General Ulysses S.: 46
Grimsley, Thornton: 38-44; *see also* Grimsley artillery saddles, Grimsley dragoon saddle, 1844
Grimsley artillery saddles: 145
Grimsley dragoon saddle, 1844: 22, 38, 51, 57, 61, 93, 145; adopted for cavalry, 41-44; replaced by McClellan saddle, 1859, 46, 63-65; used in Civil War, 46; used by First and Second Regiments of Cavalry, 1855, 50; Campbell saddle replaced by, 54

Hartman, Sergeant Henry, carbine socket design of: 94, 97
Hermès Saddlery, France: 129
Hope saddle: popularity of, 54, 57; issued to Second Cavalry companies, 1857, 57, 61; specimen of, 57; and McClellan saddle, 64-65; in Civil War, 75
Hussar-style saddles: 9, 24, 44, 65

Issue dragoon saddle, 1814: 9, 17-18

Jefferson Barracks, Mo.: 18
Jenifer, Walter H.: 75; *see also* Jenifer saddle
Jenifer saddle: 75, 78, 93
Jesup, Major General Thomas S.: 38
Johnston, General Joseph E.: 54, 57
Jones, First Lieutenant William E.: 61, 112; *see also* Jones saddle
Jones saddle, 1854: 57, 61-62, 112

Kearny, Brigadier General Stephen Watts: Grimsley saddle used by, 38; role of, in adoption of Grimsley saddle, 41-44
Kilpatrick, Major General Hugh Judson: 57

Light Dragoons, United States: in nineteenth century, 3, 9; abolished in 1816, 15; 1833 regiment of, 17-18; Second Regiment, 1836, 33

McClellan, Captain George B.: 63-64; *see also* McClellan equipment
McClellan equipment: of 1859, 22, 46, 54; adopted for United States Cavalry, 63-65, 69; modifications of, 70, 81, 85; in Civil War, 70, 73, 75, 78; issue for all regiments until 1868, 79-80; Model 1874, 89, 97; popularity decline of, 93-94, 97, 107; Model 1885, 99, 104-105, 131; Model 1904, 104-105, 115-16, 120, 131, 133, 151; Model 1928, 131-33; for field artillery teams, 145; Model 1913 mule riding saddle, 151, 155
Mackay, Colonel Aeneas: 38
Marquis Saddlery, France: 129
May, Captain Charles A.: 41
Mexican War: 41; Ringgold saddle used in, 38
Model 1832 artillery driver's saddle: 141-48, 150
Model 1832 artillery valise saddle: 142-48
Model 1917 officer's saddle: 121, 135; replaced by Phillips cross-country saddle, 1930's, 121, 124-25, 135, 137, 140
Model 1916 experimental saddle: 117, 120-21
Model 1916 officer's training saddle: 124-25
Model 1913 mule riding saddle: 151, 155
Model 1936 officer's cross-country saddle: *see* Phillips cross-country saddle
Model 1912 saddle: 108, 112-13, 115-17, 120, 124
Model 1928 McClellan saddle: in World War I, 131; still in use, 132-33; *see also* McClellan equipment

157

Model 1926 training saddle, Saumur type: 129
Mounted Rangers, 1832: 15, 17; *see also* United States Light Dragoons

Ninth Cavalry, 1866: 79

Ordnance Manual, Army's: 148

Packer's riding saddles: 155
Patton, Lieutenant George S., Jr.: 117
Phillips, Colonel Albert E.: 135, 137; *see also* Phillips cross-country saddle, Phillips packsaddle
Phillips cross-country saddle: 121, 124–25, 135, 137, 140
Phillips packsaddle: 135

Quartermaster Depot, Jeffersonville, Ind.: 140
Quartermaster Depot, Philadelphia, Pa.: 33
Quartermaster Depot, St. Louis, Mo.: 38
Quartermaster Handbook, The: 125
Quartermaster Museum, Fort Lee, Va.: 117
Quartermaster Supply Catalog: 155

Regiment of Mounted Rifles, 1846: 38, 61
Rice & Childress, San Antonio, Texas: 57
Ringgold, Major Samuel: 33, 38; *see also* Ringgold dragoon saddle, 1844
Ringgold dragoon saddle, 1844: 33–37, 43, 78; unpopularity of, 38; used in Mexican War, 38
Rock Island (Illinois) Arsenal: 108; Model 1912 saddle issued by, 112–13; Model 1916 issued by, 120
Rules for Cavalry, Hoyt's: 6

Second Regiment of Cavalry, 1855: 49–50, 54; Campbell saddles used by, 50, 54; Hope saddle issued to, 57
Second Regiment of Dragoons, 1836: 33, 41
Sedgwick, Major General John: 70
Seminole Indian Wars, Florida: 33
Seventh Cavalry, 1866: 79
Sherman, General William Tecumseh: 46, 93; Whitman saddle recommended by, 97, 99
Sixth Cavalry, 1865: 79
Stock-type saddles: 38
Stuart, J. E. B.: 41
Swords, Major Thomas: 41

Tenth Cavalry, 1866: 79
Texas saddle: *see* Hope saddle
Treatise on Artillery: 141
Turner, Captain H. S.: 41–42

United States Cavalry: 85, 112
U.S. Cavalry Journal: 108, 117, 137

Walker, James: 3; *see also* Walker contract saddle of 1812
Walker contract saddle of 1812: 3–6, 24, 28
War of 1812: 3, 9, 15
West Point Museum, New York: 33, 70, 78
Whitman saddle, 1879: 94, 97, 99